MW01064661

WANTED

WOODYS

David Fetherston

Motorbooks International
Publishers & Wholesalers

First published in 1995 by Motorbooks International Publishers & Wholesalers, PO Box 2, 729 Prospect Avenue, Osceola, WI 54020 USA

© David Fetherston , 1995

All rights reserved. With the exception of quoting brief passages for the purpose of review no part of this publication may be reproduced without prior written permission from the Publisher

Motorbooks International is a certified trademark, registered with the United States Patent Office

The information in this book is true and complete to the best of our knowledge. All recommendations are made without any guarantee on the part of the author or Publisher, who also disclaim any liability incurred in connection with the use of this data or specific details

We recognize that some words, model names and designations, for example, mentioned herein are the property of the trademark holder. We use them for identification purposes only. This is not an official publication

Motorbooks International books are also available at discounts in bulk quantity for industrial or sales-promotional use. For details write to Special Sales Manager at the Publisher's address

Library of Congress Cataloging-in-Publication Data
Fetherston, David A.
Woodys / David A. Fetherston
p. cm.
Includes index.
ISBN 0-7603-0014-3 (pbk.)
1. Automobiles--United States--History. 2. Wood in automobiles-United States--History. 3. Woodys (Automobiles)--United States--History. I. Title.
TL23.F48 1995 95-5982
629.222--dc20

***On the front cover*:** Dick DeLuna's 1936 Seabright Beach Surf Patrol Wagon. This delightful woody wagon is a special order production Dodge Westchester suburban built by U.S. Body and Forging. These cars were built on a 116in wheelbase and half-ton pickup chassis and were powered by Chrysler's L-Headed six-cylinder engine. Original price: $715.

***On the frontispiece*:** The 1948 Chrysler Town and Country automobile was the most significant wood-trimmed post-war model from Detroit. The detailing of the wood work was excellent with complex finger jointing holding the pieces together. The wooden parts were built by Pekin Wood Products in Helena, Arkansas, and then shipped to Chrysler's Jefferson Avenue plant in Detroit for assembly. The wood trim gave the Town and Country a nautical look which was carried into the chromed taillight assembly. Owner: Edwin Hawbaker

***On the title pages*:** At the Longboard Surfing Championship held at Huntington Beach Pier in Huntington Beach, California, the Southern California Chapter of the National Woodie Club held one of its regular gatherings as part of the festivities. A good showing of Fords, Buicks, Dodges and Mercurys filled the parking lot above the beach.

***On the back cover*:** One of the most elegant "wooded" station wagons of the late forties was the 1948 Packard Station Sedan designed by Al Prance and built by Briggs Manufacturing for Packard. The four-door six-passenger wagon was created by removing the roof panel and trunk lid on a Packard sedan and fabricating the wagon conversion out of ash and maple. Powered by a 130hp L-headed straight 8 engine, the Packard Station Sedan sold only to the well-heeled social set with a price tag of $3,424. Owner: Joe Scott; This '34 Ford is a perfect classic hot rod woody built not only for style but also to drive. Owner Mike DeVriendt from Colorado Springs, Colorado, built this hot rod woody in order to create a powerful hot rod with tons of space. Under the two-piece hood hides a Ford 302 V8 topped with four down-draft Weber carburetors and a matching Ford automatic transmission. Mike has won many awards for this woody, which has turned out to be one of the most popular hot rods of the year.

Printed in Hong Kong

Contents

Dedication

For Johnny Best—a woody lover, automotive illustrator, cartoonist, car nut, and loving friend . . . It's a pity we didn't know you better. . . .

Acknowledgments

My first recollection of a woody is of my father's Austin A-70 Somerset. While I can't remember much about it, I can recall his indignation at having to spend two days every spring revarnishing the wood. It faded from the family circle in the early fifties and was replaced by an uninteresting but reliable gray FX Holden GM Australian sedan. And here I am, forty years later, driving an Oldsmobile Custom Cruiser imitation woody.

My first recollection of woody wagons is of my father's A70 Somerset Austin Wagon. It's seen here towing a Carapark trailer in a photo from 1955 taken in Sydney, Australia, during a family vacation. My older brother, Michael, is sitting on the left with my grandmother standing behind, and I'm on the right. *Photo: G.H. Fetherston*

Yet the traditional woody is not forgotten. If you arrive in a real woody somewhere, a crowd will turn out like it's a million-dollar Ferrari.

The folks who own and restore these pieces of "furniture on wheels" enjoy the many woody themes. At a show you'll find a perfect 1937 120 Packard parked next to an unrestored 1951 Ford beside a 1932 hot rod woody. The woody crowd is obliging to its brethren and looks kindly on all styles of woodys.

I spent many hours researching the material I have selected for this book. I tracked down these special woodys from conversations, newspapers, and magazines so that I would have an extensive record of these great-looking vehicles. I could have pictured fewer wagons in greater detail, but I decided to show as many variations as possible.

I would like to thank the many folks who gave me their time so that I could photograph all these great woodys. I'd especially like to thank Mike Chase, who photographed Wavecrest on the left coast while I went to the right coast to cover the Vermont shows; Stan Ramondo for allowing me access to his library; Michael Lamm, Steve Anderson, and Ernst Hartley from the Towe Ford Museum; Edward Wildanger for access to the family archives; Jane Mausser, Bob Barbour, Doug and Suzy Carr, Arch Brown, and Bud Juneau for their photos and help. I also would like to thank Tony Thacker for his support with this project and Dr. Steve Werlin for handling the heart department.

I have chosen to spell woody with a "y" in this book. While the National Woodie Club spells it with "ie," I found its most common spelling uses a "y."

Woodys have become symbols of the good life. They may have started out as basic public transportation, but in their heyday they were "the wheels of the wealthy." Their decline with the coming of the all-steel station wagon eventually led to them being passed off to the "surf generation" for beach transportation, but they weren't forgotten.

The Beach Boys, Jan and Dean, and the Surfaris all sang about them, and now in the nineties they have become a small but illustrious part of modern American culture's search for the safe, fun-loving character the country possessed in the fifties.

"I loaded up my woody with the boards on top" (from "Let's go Surfing," by the Beach Boys). Long live the woody .

David Fetherston

NEW YORK

Chapter One

The Beginning

Wood has played a fascinating part in the development of the automobile. In the beginning it was an essential element for fuel, wheels, framing, and bodies. As the automobile evolved, wood became less essential. It was abandoned as a power source by 1910 but continued to be used structurally until the early fifties. From then on wood became a decorative option on station wagons, using the technology of modern plastics to replicate the look of the wooden automobile.

The evolution of horse-powered transportation, from wooden horse-drawn wagons, such as the Dutch Wagon or New England Pleasure wagon of the 1600s and 1700s to the Conestoga wagons that helped tame the wild West, was a long slow process. When civilization's search for faster and stronger transportation evolved into using steam and gasoline power at the beginning of this century, it was not surprising that the vehicles were made mostly of wood.

Our need for utilitarian vehicles rather than passenger-carrying sedans had engineers designing lightweight steel-and-wood trucks and wagons, which replaced horse-drawn express wagons.

By the beginning of the early twenties the horse had been replaced by the gasoline engine for most transportation needs. The railway spanned the country in every direction, which meant transportation was needed to and from railway depots. Horse-drawn hacks evolved into engine-powered "depot hacks."

Actually, the first recorded powered wagons were electric, but by 1910 the White Motor Company had developed the White Steam Depot Wagon, an open vehicle featuring three rows of seats similar to a horse-drawn wagon. Chase, Buick, and Pierce-Arrow also offered wooden depot hack-style wagons with wagonette tops and seating for six or eight.

With the invention of the Model-T Ford and other rapid developments in automobiles, many builders developed their own interpretations of the wagon. Virtually any good carpenter or cabinetmak-

This 1913 Ford T depot hack is typical of the body style built by Cantrell and Mifflinburg in the teens. It features a wood-framed windshield, a flat roof, and contrasting ribbed panels built of oak and cherry.
Owner: Vince Iaccino

er could build a wagon body.

The wooden station wagon, as we know it, came out of this period. Manufacturers sold rolling chassis that allowed body builders to install their own custom-built bodies. Companies such as Post in Farmingdale, New York; Parry in Indianapolis, Indiana; Hoover in York, Pennsylvania; and J.T. Cantrell in Huntington, New York, jumped onto the bandwagon, so to speak. Initially these new wagons were called Suburbans, Combinations, or Country Clubs, but really they were all depot hacks. They featured wooden structures with wagonette-style tops.

The Hoover Wagon Company produced one of the first wooden station wagons in 1914. This body was fitted to a standard Model T Ford and could be adapted to any rolling chassis combination. It featured a full top with side curtains, two rows of seats, and a side door, along with some of the first wood panel and rib construction.

Until this time most of the wood panels were actually planking, but builders began searching for lighter construction methods. The results were ribbed and

1910 Lotis Estate Cart

An early British woody, the Lotis "estate cart" had folding seats and removable cushions, forming either a ten-seater shooting brake or a medium-weight wagon.

1911 Champion Electric Depot Hack

By 1911, the first of the rib-and-panel woody bodies emerged. Champion Electric offered a depot hack with a multiple seating configuration based on a wagonette body. The driver sat exposed up front using a vertical steering column.

framed side paneling.

As the twenties approached, the number of body builders who sold suburbans or depot hacks increased. Among the major players were Hatfield of Sidney, New York; Mifflinburg Body Company of Mifflinburg, Pennsylvania; Columbia Body Corporation of Columbia, Pennsylvania; and the York Body Corporation of York, Pennsylvania. There were also manufacturers outside the northeastern states building these kinds of bodies, including the Eagle Pass Lumber Company in Eagle Pass, Texas; Stoughton Wagon Company of Stoughton, Wisconsin; and Larkins and Company in San Francisco, California, which had built stage coaches for Wells Fargo.

These companies built wagon bodies that suited the transportation needs of families and the public but also found use on ranches and with small businesses as produce and goods haulers. They were fitted to every kind of vehicle, but the most popular and enduring was the Model T Ford.

In these times a customer could order a chassis from their local Ford dealer. The chassis would either be shipped to the body builder or the body would be shipped to the dealer for installation. Thousands of vehicles were ordered in this manner. One of the biggest suppliers was Martin-Parry who owned factories in Indianapolis, York, Pennsylvania, and Lumberton, Mississippi. Martin-Parry manufactured bodies and kits at its factories and also serviced over fifty assembly plants across the country, laying claim to the title of "The Largest Commercial Car Builder in the World."

Chapter Two

The Twenties

Chevrolet entered the wagon business briefly in 1920. The company listed a "Light Delivery Wagon" for $770 built at its Flint, Michigan, plant. The wagon apparently didn't get the response the company wanted as it rapidly disappeared from the sales listings. By this time there were dozens of body builders across the nation producing station wagons for almost every make of chassis.

However, a change was in the wind. A mahogany-paneled woody station wagon was built by Healy and Company in New York to a design prepared by J.R. McLauchlen of Cadillac custom body. It was mounted on a 1922 Cadillac chassis and featured stylish wood paneling and ribbing, whitewall tires, and three rows of seats. It was the hit of the social season on the Jersey coast in 1923. This Cadillac woody lifted the status of the lowly depot hack from a simple method of transportation to a traveling social statement.

By the mid-1920s the aftermarket for wood-bodied suburban/depot hack-style wagons had expanded to dozens of new manufacturers; yet, none of Detroit's automobile or truck manufacturers were offering a production station wagon body.

The Durant Motor Company revived the Star automobile in 1922 when W.C. Durant, the former head of General Motors, purchased the Star Motor Company. One of his first objectives was to set his company apart from the others in the market with a "production" depot hack. Durant had the common sense to market a new vehicle that he didn't have to build. He had the Stoughton Wagon Company and Martin-Parry supply him with bodies that he fitted at the factory and then delivered as finished wagons to his dealers.

Buick had the same idea and introduced its Combination Passenger and Express wagon later

This 1926 Ford Model T was built by Michael Macica to replicate the depot hack theme of the mid-1920s. The wood craftsmanship is stunning with its oak paneling and cherry wood trim contrasting superbly against the red/maroon and black paintwork. The red painted wire wheels add more interest, as does the side-mounted wire wheel. *Owner: Michael Macica*

The Canopy Express was another variation on the wood-bodied truck/depot hack concept. This 1921 Ford is powered by the traditional four-cylinder T engine, rolls on wood-spoked wheels, and features an oak and cherry wood body built by Babcock Body Works of Watertown, New York. *Owner: Carl Goodman*

that year. The Buick, which sold for $935, featured a body much like Joseph Wildanger had just started building in New Jersey. It had flat side panels featuring fine ribbed bodywork that fit inside the rear fender line. Unlike the Star, Buick offered a three-model line-up of wagons.

By the mid-1920s body builders had developed special features to offer their customers. Columbia Body had ventilator-type side curtains, removable rear seats, linoleum floor coverings, and rubber toe kicks. The Waterloo Body Corporation from upstate New York had emerged from the Waterloo Wagon Company around

1920 after making horse-drawn wagons for forty years. Their specialty was patented flexible seating arrangements for their station wagon bodies specially designed for Ford, Chevrolet, Dodge, and Willys-Overland chassis.

Another variation came from the Cotton Beverly Company of Concord, New Hampshire. Their wagons used production chassis, but instead of production cowls, they built extended cowls covered in imitation leather which provided front seat passengers with far more leg room. The rest of the body was built up using

By the early twenties, thousands of depot hacks had been built and installed by outside suppliers on bare chassis from virtually every manufacturer. However, in 1922, Billy Durant, head of the Star Motor Company, saw an opportunity to offer a factory-installed depot hack to his customers. The Stoughton Wagon Company-supplied body could be ordered with extra seating for up to seven passengers and with such luxuries as roll-down side curtains. Durant was not alone in this trend-setting move; Buick offered a similar Combination Express Wagon in 1922 for $935. *Photo: Arch Brown Photo Collection*

matching imitation leather covering with natural wood-finished white ash moldings. This style of woody body was also seen on larger, heavier chassis from the Dodge Brothers, Buick, Franklin, and Essex.

The manufacturers of these wood-framed and trimmed bodies described their products as wagons or suburbans, and most of them were sold for daily transportation needs and hauling.

Meanwhile, there was a clientele looking for some fancier versions to use on the estates and country clubs of Long Island, the Jersey coast, the Upper Peninsula of Michigan, and the Catskills. These wagons were the perfect foil for the social set, offering custom styles that carried their estate or club name embossed on the door, like traveling royalty.

Offerings from Mifflinburg, J.H. Mount, Columbia Better Bodies, Cantrell, and others expanded with seating for seven, full doors, and better seat cushioning stuffed with "curled hair." Builders developed their own touches, special ironwork or woodworking that identified their style of body.

By this time Mifflinburg Body Company had become one of the major players in this market, employing over three hundred people and producing sales of over a million dollars.

Joseph Wildanger was one of these body builders who had his own "wooden signatures." Wildanger held a meisterschein, or master certificate, in carriage making, earned while he worked as an apprentice, and then journeyman in Vienna, Austria, for seven years.

With his young wife, Wildanger arrived in New York around 1910 looking for a better life, and he quickly found work in the auto body building business. His

The classic 1924 Hispano-Suiza H6C is considered one of the world's greatest race cars. With its lightweight tulipwood body, this racer was one of the most technically advanced cars of the twenties, featuring a top speed of over 110mph, using a 200hp Boulogne, six-cylinder engine and four-wheel brakes. It was built for André Dubonnet, a World War I fighter ace and heir to the French aperitif company fortune, who raced it in such great events as the Targa Florio. The Hispano-Suiza's wonderful wooden bodywork was made by the famous Nieuport aircraft company using strips of machined tulipwood fastened in place with masses of evenly spaced double copper rivets. The vertical rivetting was so exact that the fender rivetting matched the rivet lines in the body. The body was trimmed with matching copper sheathing which created a polished bellypan; copper plating was also used extensively on the wire wheels and trim. *Photo and Owner: Behring Auto Museum, Danville, California*

This 1924 Dodge is a perfect example of what a person can do with a chassis, some fenders, a cowl, and a good imagination. Mike Chiavetta built this replica woody using a set of plans he obtained from the National Woodie Club. Based on a Dodge chassis, Mike scratch-built the body out of maple with birch plywood paneling and basswood for the headliner. It is powered by a 350 Chevrolet and rolls on American Racing alloy wheels. Mike is also a mural artist and is responsible for the woody mural on the beachfront in Huntington Beach, California. *Owner: Mike Chiavetta*

skills in carriage making soon had him working as shop foreman for J.H. Mount, building wagons and truck bodies, but he held higher aspirations and eventually set up his own business, Jos Wildanger Co, in 1922 in Red Bank, New Jersey.

Wildanger's suburban bodies featured an interesting ribbed paneling style that set them apart from the builders. His design used a series of fine horizontal maple ribs set about 6in apart over either wooden or painted metal panels with all the bodywork set within the confines of the rear fender wells.

The company flourished and expanded from making wagons to truck bodies, police paddy wagons, armored cars, and ambulances. Company records show over five hundred wooden wagons were built before 1942, when the company closed up shop for the duration of the war. In the postwar years, Wildanger converted to commercial vehicles, building funeral vehicles along with vans and truck bodies.

For many of the builders manufacturing truck bodies, the wagons were a kind of icing on the cake that gave them a second string to their production flow.

With the introduction of the Model A Ford in 1928 came a new generation of basic transportation that suited the woody wagon builders just fine. Then Henry Ford de-

Jos Wildanger Company was one of the many small East Coast body shops that built commercial truck and wagon bodies in the late 'teens and early twenties. Based in Red Bank, New Jersey, Wildanger's shop on West Street built hundreds of wagons on all kinds of truck and passenger-car chassis. This 1925 Studebaker displays Wildanger's unique ribbing style, which fit multiple horizontal ribs over painted sheet metal. *Photo: Edward G. Wildanger Archives*

The British had their uses for woody wagons, too. One of the most interesting aspects of British shooting brake history revolves around the rebody business. Many shooting brakes were rebodied luxury sedans and limousines. This 1926 Rolls Royce was originally fitted with a Windovers limousine body. Powered by a 20hp six-cylinder Rolls Royce engine, it was rebodied at the end of World War II with this stylish ash shooting brake body. It rolls along on Rudge Whitworth wire wheels and features sleek rear fender skirts with a center cap cut-out. *Photo and owner: David Dodge*

cided to use Billy Durant's idea of a wagon as a production model, and on April 25, 1929, the first of 4,954 Model A Ford station wagons rolled off the production line.

The maple and birch bodies were built by Briggs and the Murray Corporation in Detroit from preformed timber components supplied by the Mengel Company, a furniture manufacturer in Louisville, Kentucky.

The styling did not progress as dramatically on wood station wagons as it did on the new all-steel body sedans; the engineering for these wagon bodies still used handcrafts carried over from the days of the horse-drawn wagon.

This is another rebody of a Rolls Royce. This 1927 20hp was supplied to the Earl of Moray fitted with a touring body and 21in Rudge-Whitworth wire wheels. In the late forties it was rebodied in oak by A.R. M. Cleod and Sons, a boat-building company, for use on a Scottish estate. The current owner purchased it in 1957 and uses it as a family touring wagon and camper. It was recently restored and can be seen here, outside Blenheim Palace in England. *Photo and owner: David Mitchell*

This 1926 Dodge Brothers four-cylinder Estate Wagon was created out of a sedan which had already been converted to a truck back in the thirties by the owner's grandfather. It had been parked and abandoned on the family ranch in San Jose, California, in 1943 and the late-1970s, Dave Chiotti decided the old Dodge would make a great woody project. He started by restoring the running gear, chassis, and sheet metal before crafting the estate wagon body out of oak over the following eight years. The timber is stained with a walnut finish and Dave had Michael Parodi in Sebastopol, California, paint the sheetmetal in black and brown with orange on the disk wheels. Dave's design was based on period advertising images for estate wagon bodies offered by Dodge. It features two rows of seats with the rear unit folding a per the original model. Dave researched the Dodge's production and found it was manufactured in June 1926 as one of the first 6-volt models which still used the old backwards transmission pattern. *Owners: Dave and Marilyn Chiotti*

Like many hot rod woodys this 1928 Ford features an owner-built maple body. The wagon began as a simple chassis and cowl, with a 400ci Chevrolet V-8 producing the horses. Paul Dunne then fabricated the body from the floor up. His work included the nicely rolled top, a visor, and building the seats. *Owner: Paul Dunne*

The Ford sales brochure in 1929 listed the Station Wagon: "It can be used for passengers or as a delivery unit and makes a most valuable addition to country clubs and estates."

The Station Wagon

THE new Ford Station Wagon, used either as a passenger carrying car or a delivery unit, makes a most valuable addition to the motor vehicle equipment of country clubs and estates.

As a passenger conveyance it accommodates eight persons, with ample space for luggage on the lowered tail gate, if required. Rear seats can be removed to convert the car into a haulage unit.

The body combines attractiveness with rugged construction. It is finished in natural wood, hard maple being used throughout.

A wide seat in the driver's compartment will accommodate three persons. In the rear compartment with two single seats and a full-width seat, accommodates five persons. All the seats are deeply cushioned and upholstered in blue-gray artificial leather of Colonial grain. Rubber floor mats are provided in both compartments.

The roof, supported by hard maple up-rights, is covered with heavy black deck material.

Side curtains are of rubber-interlined material, tan-gray in color to harmonize with the body finish.

Dimensions are as follows:

WIDTH OF FRONT SEAT	45	inches
WIDTH OF REAR SEATS		
2 Single	18	inches
1	44	inches
WIDTH OF DOORS	27½	inches
LOADING SPACE		
Length	59	inches
Height	48	inches
Width	50	inches

Passenger Compartment

INTERNATIONAL

Chapter Three

The Thirties

The thirties got off to a deadly start. The stock market crash of 1929 struck a near-fatal blow to much of the auto industry. Not only did many of the minor players in the auto-building business close up shop permanently, but thousands of businesses in the aftermarket were forced to call it quits.

Huge body builders like Mifflinburg and Martin-Parry were devastated by the Wall Street crash and the popularity of Ford's own station wagons. Martin-Parry sold off most of its production facilities to Chevrolet at the end of 1932, while Mifflinburg scaled down its station wagon shop and focused on building truck bodies. Mifflinburg advertised that they were still in the station wagon business offering a selection of custom built wagon bodies ranging in price from $325 to $350. The bodies had a shipping weight of 1,100 to 1,200lb and could be ordered with either storm curtains or roll up glass in all doors. Most of Mifflinburg's bodies were fairly utilitarian with square-edged bodywork and little styling to improve their lines.

They did, nonetheless, make one breakthrough in body construction in the mid-thirties; they offered the first of the all-steel station wagon bodies built on a Chevrolet half ton chassis. Unfortunately, it failed to make much impact in the market place. (Mifflinburg eventually closed up for good in 1941.) Some of the smaller shops like Wildanger survived doing repairs and commercial work along with a limited amount of custom wagon building.

Others who survived the crash included the York-Hoover Body Corporation from York, Pennsylvania. The corporation had evolved from a merger of the York and Hoover body builders, but they only managed to struggle into the thirties. York-Hoover offered the "Estate Suburban Body" which was noted

Even International Harvester got into the woody wagon business. International used an outside supplier for all body conversions on its C-1 light truck chassis. When the National Park Service ordered seven in 1938, the bodywork was turned over to M.P. Moller in Maryland. This wagon saw service at the Ahwahnee Hotel in Yosemite National Park, California, for guest transportation. Painted tan with green wheels, it was created from ash, oak, cottonwood, and mahogany. *Owner: Bruce Campbell*

Right: This 1930 Franklin 147 wagon is a wonderful example of the Cantrell body during this time. This body was replicated by George Clapp in ash and basswood. The big six-cylinder air-cooled engine is smooth running, and the Franklin rolls along on wooden-spoked Motorwheels. *Owner: Henry D. Maywell*

The small wagon and commercial body-building companies had their own special design details. Joseph Wildanger bodies featured ash framing with delightful fluted horizontal ribbing fastened to painted metal panels. This 1930 Wildanger Ford Estate wagon was built by the third son of Joseph Wildanger using a chassis from a two-door sedan, company design papers, photos, and some full-scale drawings he created with the help of his brother Arthur, who had run the Wildanger company for sixty years. The early Wildanger bodies were clean, straight-line designs that featured all the bodywork between the rear fenders and only a couple of curved pieces in the roof. *Owner: Edward Wildanger*

By the early thirties the Jos Wildanger Company was well known on the Jersey Coast for its quality custom-built estate wagons. Despite the Wall Street crash and a close call with bankruptcy, Wildanger managed to survive, building a handful of wagons and commercial vehicles along with all kinds of woodwork and repairs. Like the earlier Wildanger-bodied Studebaker wagon in the book, this 1932 Pontiac Series 90 wagon features the interesting Wildanger ribbing, which used tapered edges with squared-off ends fitted over painted sheet metal, with the wagon body fitting between the fender wells. Pull-down roller curtains were one of the options on this wagon. *Photo: Edward G. Wildanger Archives*

in their ad as "useful for the countless summer tasks of the country estate, or for the carrying of equipment to that favorite fishing hole or bathing beach."

Hercules Products Inc. from Evansville, Indiana, builders of Hercules bodies, managed to survive the crash but only just. The company's line of Better Business Bodies for Chevrolet chassis were similar to others in the field but they also offered a more attractive line of "Aristocrat" bodies especially designed for the half-ton Chevrolet commercial chassis. These bodies featured glazed windows and full-skirted side panels with a stylish "beaver-tail" rear end.

Hercules noted that their wagons were "Ideally Suited to a Wide Range of Commercial and Pleasure Uses" and to "Forget the week-day cares, take along the

By 1931 the style and performance of Ford's new Model A had made it into an amazingly big hit even though the country was struggling to escape from the Depression. This beautifully restored Manila Brown Station Wagon features a Raulang oak and birch body, 19in wheels, side curtains, and a radiator guard. *Owner: Bill Mohler*

This 1932 Ford woody is the work of two of the finest automotive designers and craftsmen in California. Designed by Thom Taylor as the kind of woody he would build for himself, this wagon was eventually built by Dan Fink of Dan Fink Metalworks in Huntington Beach, California. Painted in Ford Blue and powered by a 285hp 5 liter V-8 with GT-40 heads and fuel injection, it rolls on a complete Pete and Jake chassis. The superb woodwork was done by Ron Heiden, using a complete steel frame wagon body as support for the handcrafted framing and paneling. *Owner: Dan Fink*

picnic things, there's plenty of room."

The Hercules Standard Model 50C1 was the company's general-purpose model and was sold as a utility vehicle for local transportation services to resort hotels, country clubs, camps, airports, county surveyors, engineers, and builders.

Times were tough and in order to survive Hercules merged with the Campbell Body Company to form Hercules-Campbell Body Company, Inc. The operations were moved to Tarrytown, New York, where the company continued to build wagons, focusing mostly on bodies for Chevrolet commercial chassis.

Four lines were offered: a general Purpose Station Wagon, a DeLuxe Station Wagon, a Suburban, and an Enclosed Suburban. The Suburban models were stylish with rib and panel construction, featuring molded ribs and a similar flowing "beaver-tailed" rear end treatment to the older Hercules Aristocrat line.

This beautiful stock-bodied 1932 Ford Model B features a Raulang body restoration done by Bill Larson of Elco, New York, using traditional maple frames and birch panels. As a classic color contrast the body is finished in Desert Tan with black fenders and rolls on wire spoke K.H. wheels. Under the hood hides a four-cylinder engine running a center weighted crank and a Miller overhead valve conversion. *Owners: Mr. and Mrs. Fred P. Montanari*

Apart from the few manufacturers who did manage to carry on building aftermarket station wagons, Ford was now virtually the only source of production station wagons. Henry Ford continued to contract with the Mengel Company to build wagon parts, which were then shipped to Murray, Briggs, and Baker-Raulang for assembly.

Baker-Raulang of Cleveland, Ohio, evolved from the electric automobile manufacturers Baker, Rauch, and Lang, who had built electric cars from 1899 through 1928. First named Baker-Raulang, the company later evolved into Baker-Rawling, and by the late twenties its main source of income was body building for Ford and International.

Henry Ford had the foresight to purchase the town of Pequaming and the half-million acres of hardwood forests surrounding it on Michigan's Upper Peninsula. From these forests came the maple framing and birch paneling for Ford's new station wagon.

Ford had been buying lumber from the mills on the peninsula for the framing on his Model Ts, but he opened his own Iron Mountain sawmills in about 1922 under the Michigan Land, Lumber and Iron Company in the Menominee River Valley. This operation quickly expanded to three lumber mills and a waste wood conver-

The advent of the flathead Ford V-8 trumpeted a new era in power and performance, and the styling of the 1933 made it an instant classic. In keeping with its theme, this 1934 wagon was also built to drive. Powered by a Ford 302 V-8 and topped with four downdraft Weber carburetors, it has plenty of hauling power to move it through the air. *Owner: Mike DeVriendt*

sion facility that turned out charcoal briquettes for home heating.

Along with the mills, Ford opened iron and copper smelters using hydro-electric power to convert raw materials into production metals. This was not only cost effective at the time, but it allowed him to control the flow of raw materials to his other factories.

At first the Iron Mountain mills turned out raw lumber. Then a staff of woodworkers made preformed wood components which were shipped to Briggs and Murray in Detroit and later to Raulang in Cleveland for assembly. This system changed when Ford's own station wagon assembly line opened in the fall of 1936 at Iron Mountain.

By the mid-thirties Wildanger bodies took on a more conventional look. They dropped the delightful ribbing style that had been a company touch throughout the twenties. This 1934 Buick conversion features a fully glazed body with ash framing over maple paneling. Simple flat spring steel retainers kept the back door and lift-up window secured. The construction of these bodies was straightforward; the only curved woodwork was in the roof line and lower door frame. *Photo: Edward G. Wildanger Archives*

Below: This 1934 Chevrolet station wagon was the last custom-ordered woody built by Larkins and Company before they closed shop in late 1934. Delivered to the Fleisbacker family in San Francisco, the ash-bodied wagon was chauffeur-driven from 1935 to 1946 for outdoor-painting trips into the country. After World War II, it was shipped back to Maine, where it remained in storage until 1972. It was eventually sold to Peter Budd in Ontario, Canada, who carried out this stunning restoration on the six-cylinder wagon. It may be one of a kind; the recent owners have not traced any others from this builder. *Owner: Larry Zink*

The assembly line allowed Ford to fabricate and assemble at one site. This not only lowered the cost of shipping but improved quality; the workers could see the fit and finish of the bodies as they came off the production line.

As the wagons were finished they were moved onto rail cars, ten at a time, and shipped to fifteen Ford assembly plants all over the country.

Initially wood was used in all of Ford's production automobiles, but as automobile body design moved ahead, the need for wood products eased off leaving Ford with a mass of hardwood, which suited wagon production perfectly. These wagons came with side curtains fitted with transparent plastic isinglass windows; however, this product did not have a long life, as many owners discovered. It scratched easily, fogged up from age, and became brittle in cold weather.

In 1936, glass was fitted to the front doors of Ford wagons, and in 1937 it was fitted throughout. Chrysler followed suit in 1937, offering full glazing.

Many station wagon bodies looked similar, yet there were some interesting trimming and style differences among them. The Ford bodies had their own distinctive mix of vertical and horizontal ribbing that criss-crossed the panels between the frames. Others, including Cantrell, de-

This 1935 Pontiac was another wagon built by the Jos Wildanger Company in Red Bank, New Jersey. Like many of Wildanger's earlier wagons, it features wood framing with painted sheet metal paneling. Joseph Wildanger came to the United States from Hungary just after the beginning of the century. He was a trained carriage builder and worked for the body builder J.H. Mount in New York until he opened his own shop in 1922. His sons later took over the company and ran it for the next sixty years. *Photo: Edward G. Wildanger Archives*

The Europeans were enjoying wood styling, too. This 1935 Franay-built Hispano Suiza is one of 206 Type K6s, built between 1934 and 1937. Powered by a large and powerful six-cylinder 5 liter engine, this wagon version is one of only a few fitted with shooting brake bodies. It is seen on display at Retromobile in Paris, Europe's greatest car collector show.

The British followed the American utility wagon concept. This 1935 20/25hp Rolls Royce, powered by its original 4liter six-cylinder engine, was converted in 1950 by Lieutenant Colonel J. Anderson in Inverness, Scotland, to this shooting brake body. It features a pair of folding sleeping bunks and was modified to pull a matching, custom built camping trailer. *Photo and owner: Robert Storey*

veloped a mixture of ribbing styles, while U.S. Body and Forging used mostly structural framing and plain paneling.

By the mid-1930s what was left of the independent body builders could be counted on two hands. The Hercules-Campbell Body Company from Tarrytown, New York, merged with the Waterloo Body Company in New York to form Mid-State Body. Robert Campbell took over as president, and by the end of the decade, Mid-State was building most of Chevrolet's wagon bodies.

It was not just The Big Three who were turning out wood station wagons. They could be ordered from Willys, Hupp, Graham, Hudson, Packard, American Bantam, and even your friendly Studebaker dealer. Interestingly, Studebaker had started in the wooden vehicle business nearly a century before, building wooden wheelbarrows.

Chrysler continued dabbling in the wagon market using bodies built by U.S. Body and Forging in Tell City, Indiana. Finished to a higher level of trim and quality than the Ford bodies, Chrysler wagons featured trimmed door panels, piano hinged doors, and rain gutters. They were built with white ash framing and maple paneling and trimmed with a distinctive red gum strip just below the window level.

Even though the Depression had closed most of the small body builders, the best survived through good luck and good management—and accepting woodworking assignments ranging from coffins to furniture and pianos. Cantrell was one of the lucky ones and continued building bodies for Chrysler and GM chassis, although on a very restricted scale.

Cantrell offered a Dodge-based wagon called the Cantrell Carryall in 1935, while at the same time Plymouth and Dodge offered their new six-cylinder PJ Deluxe Westchester Suburban wagons, built by the U.S. Body and Forging Company in Buffalo, New York. This body style continued until late 1938 on the half-ton Plymouth and Dodge pickup chassis.

In these fickle times, Cantrell and Wildanger did what they could to survive, maintaining "estate wagons" and building some new wagons for an exclusive clientele on the Jersey coast.

Packard was among Cantrell's clients, and in 1937 Packard introduced its first production wood station wagon with a Cantrell body. Two models were offered, one on the famous 120 chassis and the other based on the new 115-C chassis.

The station wagon's newfound status came with a price that put it out of the reach of the common folk. Its popularity with hotels and private resorts for guest transportation and with the wealthy for transportation to and from their private estates was, in a way, a return to its utilitarian roots when wagons were considered "depot hacks."

Not to be out of pace with the market, Pontiac offered its first production wagon model in 1937 based on its six-cylinder passenger-car chassis. Bodies were supplied by Hercules in Henderson, Kentucky, and featured maple framing with mahogany paneling, safety glass in the front doors, three rows of seats, and Pyralin-windowed side curtains.

In 1939, Pontiac changed to fully glazed bodies supplied by Ionia and Hercules. The bodies were similarly styled; but the Hercules body used white ash frames with mahogany paneling while the Ionia body used white ash with birch paneling and more ribbing.

The year 1934 was very successful for Ford, but it was also the end of the company's two-year design cycle. The 1935 model took the good points and improved upon them with a few styling and detail changes, a new chassis. The wood detail on this 1935 wagon body shows the maple framing and ribbing that denoted Ford's station wagons. *Owner: G. West*

The maple frames and birch panels of this Iron Mountain-style 1935 richly contrast the beige body and Poppy Red wire-wheels capped with nonproduction wide-whites. This wagon features full-glazing, external horns, and a single windshield wiper. *Owner: Joseph Bamford*

Studebaker also entered the wagon business around this time with the Suburban Car using bodies supplied by the U.S. Body and Forging Company and featuring fully glazed windows. The wagons were offered in 1937 until 1939. By 1940, Studebaker was out of the wagon business.

In Europe the woody was called an "estate car" or "shooting brake." These vehicles looked like the American station wagon but were custom built for wealthy owners who could afford such luxuries. Models were built on production chassis from Rolls Royce, Bentley, Wolseley, Delahaye, and Hispano-Suiza.

Other manufacturers were in the wagon business, although on a small scale. International Harvester offered a station wagon based on its midsized C-1 truck chassis. The company used the services of three body builders over the years: M.P. Moller Inc., of Hagerstown, Maryland; Baker-Raulang, of Cleveland; and Burkett, of Dayton, Ohio.

These International Harvester wagons served in many capacities, from jitneys to school buses to family transportation. Willys also entered the market with a U.S. Body and Forging wagon which they offered for two years virtually unchanged.

Some manufacturers exported bare commercial chassis with wagon kits ready for assembly. These were shipped to places as far afield as Australia, India, and England. In England dealers sent bare chassis to body builders to be outfitted. Many one-ton Dodge chassis were shipped and furnished in this manner.

As the decade came to a close, the new Ionia Manufacturing Company now supplied bodies to Chevrolet, Pontiac, Oldsmobile, and Buick. Ionia evolved from the Ypsilanti Furniture Company, which Don Mitchell had purchased on the brink of bankruptcy in 1938 and renamed Ionia Manufacturing.

Custom-built sedan bodies were a common practice among the wealthy in the mid-1930s. While some were wildly extravagant, others were simple and tasteful, such as this beautiful 1936 Ford sedan, apparently built for Harry Bennett, Henry Ford's right-hand man. Based on a Deluxe four-door sedan, the body has been stretched and the rear section is all-new fabrication. It is painted in Washington Blue, which contrasts with the delightful oak framing and maple paneling. *Owner: Dennis J. Gallagher*

Following pages: In 1936 U.S. Body and Forging did these very tidy Westchester Suburban bodies on Dodge and Plymouth half-ton pickup chassis. This Dodge was built with kiln-dried white ash framing and maple paneling with a red gum strip at the belt line. Inside it features heavy-duty tongue-and-groove flooring and seating for eight. It is set up as the Seabright Beach Santa Cruz Life Guard Patrol wagon with all the official trimmings. The Dodge was part of the Harrah's Reno, Nevada, auto collection from 1976 to 1986. *Owner: Dick DeLuna*

LIFEGUARD
SEABRIGHT BEACH

UN-OFFICAL USE ONLY

CALIFORNIA
SURFN 37

Left: In 1936 Ford once again revised its line with a new body and grille. The horns were now hidden, and steel drop center rims featuring chrome hub caps replaced the wire wheels in the year of the three-millionth Ford V-8. This wonderfully preserved Station Wagon features restored factory-original wood with new Washington Blue bodywork and refreshed V-8 running gear. *Owner: Thomas H. Doyle*

Left bottom: In 1937, Ford's Station Wagon came with roll-down windows in the front doors only; however, the wagon could be fully glazed for an extra $20. Dealer stock price for the wagon in 1937 was $775. This maple and mahogany woody hot rod carries forward the tradition but not the running gear. It was owner-built and is powered with a 350 Chevrolet, Turbo 350 transmission, and rolls on wire wheels. *Owner: Sharyn Legg*

By the mid-1930s Chrysler was turning out quite a selection of woody wagons. De Soto, Chrysler, and Plymouth all had their own versions. This original 1937 Plymouth features a 210ci factory six-cylinder flathead with a U.S. Body and Forging-built commercial body. This unrestored original Plymouth features an oak-framed body with a gum wood strip at the belt line over factory tan paint. *Owner: Lee Kidwell*

DIMENSIONS OF LOAD SPACE: With both seats removed, load space is 85½ by 56 inches; from dash to tail gate, 134¼. Aisle at floor, 13½ inches.

A station wagon, by its very nature, should be a *smart-looking* car . . . and this Terraplane for 1937 is about as stylish a turnout as ever appeared on the highway! More than that, it has ample room to carry eight persons in complete comfort, with space on the tail gate for luggage.

Economical in cost and in operation, this Terraplane is perfectly adapted to serve hotels, resorts, country clubs, as well as private estates, schools, airports, sanitariums, road construction engineers, surveying crews, etc., etc.

The centre and rear seats are easily removed whenever it is desired to use the whole space for baggage or other articles. The end gate opens downward and can be used to carry additional baggage. Tail light is recessed.

Terraplane also got into the Station Wagon business briefly, as this 1937 ad displays so nicely. *Ad from the collection of J.B. Donaldson*

Today any Packard woody wagon is considered a "jeweled classic." Always known for style, power, and performance, Packard liked to deliver the kind of custom car the wealthy sought. This nine-seat 1938 Packard Model 1900 six-cylinder wagon is stunningly restored and still driven on tours and to shows. Its dark green paint and perfect ash-and-mahogany body beautifully reflect the crisp style that Packard set from nose to tailgate. *Owner: Tom Daly*

The changes from 1937 to 1938 on the Ford line-up were mostly cosmetic. The base model now became the Standard with a different hood and wheels from the Deluxe model and an 85hp V-8 engine, which offered plenty of power. These were Iron Mountain wagons, and for the first time, they featured glass as standard, with wind-up glass in the front doors and sliding or fixed glass in the others. The body on this beautifully restored 1938 was built with maple and birch. *Owner: R.C. Swanson*

Left top: This most unusual 1938 Wildanger-bodied Oldsmobile was obviously built on a passenger-car chassis. The shorter chassis with its tall windshield didn't allow the designers as much space as they had with the commercial chassis on which most wagons were built. The shortness of the doors and the rake of the roof give this wagon quite a different look. *Photo: Edward Wildanger Archives*

Left bottom: For the second time in its history, Chevrolet offered a station wagon as a production model in 1939 with the Master Deluxe and the Master 85 wagons. Even though these Chevrolets were priced almost $100 less than similar Fords and powered by a six-cylinder 85hp engine, the wagons did not sell in great numbers. This Campbell-built ash-and-mahogany wagon offers seating for eight and features original woodwork. *Owner: T.R. Holz*

'42

Left: In 1939, Pontiac sourced its bodies from several companies, including Ionia and Hercules. Powered by a flathead six-cylinder 80hp engine, this Hercules-built Quality Six Pontiac wagon was structured in white ash with mahogany paneling. The wagons all featured seating for eight, full glazing, and rear opening doors.
Owner: Robert Stevens

Only the front sheet metal was changed on the 1939 Ford Station Wagon. The body remained the same as the 1938. Standard and Deluxe models were offered, but the doors still opened at the center post. This hot rod restoration is a daily driver and has been extensively revised with a stronger chassis, disc brakes, and a fuel-injected '87 Pontiac Trans-Am 305 V-8 with a Turbo 350 transmission. Finished in PPG airport turquoise and rolling on Halibrand-style wheels, it's a perfect hot rod woody. *Owner: Bud Root*

CALIFORNIA
RDMSTER

Chapter Four

The Forties

As the war approached, the market was doing quite well selling a wide selection of wagons to upscale clients. Plymouth introduced its new passenger chassis version in 1940, assembled using components supplied by Pekin Wood Products and U.S. Body and Forging.

Ford continued to dominate the market with its crisp new 1940 model. Gone were the suicide doors, which were replaced with front-hinged units featuring full glazing. Ford delivered close to 10,000 Standard and Deluxe station wagons that year.

Chrysler arrived on the market in March 1941 with a new Town and Country series. Its station wagon styling featured a rounded "barrel-back," which was structurally framed in white ash with molded Honduran mahogany paneling.

This new Chrysler was quite a departure from the station wagon concept. Its four-door configuration was designed to carry up to nine passengers with a huge trunk, steel roof, full glass, and all the comforts of a regular sedan.

Before the start of the war Chrysler managed to build just 1,995 of these beauties. They used Pekin Wood Products in Helena, Arkansas, to make the components for these bodies based on a concept that had apparently come from Paul Hafer of the Boyertown, Pennsylvania, Body Works, via a proposal to Dodge.

David A. Wallace, president of Chrysler, liked the idea and took the concept, turning it into a production reality. Wallace was also president of Pekin Wood Products, so using them to supply the components simplified the process.

This was the year that Ford introduced a Mercury version of its station wagon with its new 1941 body series, which eliminated the running

Buick also continued as a player in the woody market. This superb 1948 Roadmaster Estate Wagon came with Dynaflow automatic transmission, was sold as an exclusive model, and was priced accordingly higher than the regular Buick line-up. With only 33,000 miles on the clock, this completely original Roadmaster shows how durable wood wagons can be when they have been well maintained. *Photo and owner: Bud Juneau* boards. All Ford bodies were now built at the Iron Mountain plant using timber harvested from Ford's own forests. These wagons were still structurally framed in white maple while the paneling was done in either gum,

Right: The Packard station wagon was the ultimate in rolling social statements. It exuded power, class, and style, and its 120hp straight-eight gave it plenty of go. Designed as an eight-passenger, this wagon was built by Hercules after Packard switched from Cantrell at the beginning of 1940. This 120 was totally restored in ash in the mid-eighties and had once belonged to S. S. Kresge, the founder of both Kresge's department stores and K Mart chain of stores. *Owner: P. Montano*

Below: Ford continued its two-model line-up of wagons with the Standard and Deluxe in 1940. New features included front hinged doors and simplified ribbing. This hot rod version has been revised with a 305 Chevrolet V-8, power steering and brakes, and Mustang II front suspension, and features a maple body by Ron Heiden. It carries the neat license plate "WZATREE" and tows a matching 1948 teardrop "woody" trailer made of laminated maple and mahogany. *Owner: Cliff Parker*

Right bottom: Willys ventured into and out of the woody market on several occasions. At the beginning of the 1940s the company offered a limited run of wagons called the Town and Country. As few as five may have been built. The bodies were apparently supplied by U.S. Body and Forging. This 1940 was purchased in San Francisco with the original bill of sale, unrestored, with 52,000 miles on the clock. *Owner: Zane Cullen*

NEW YORK
9965198

One of the most sought-after woody wagons is the 1939-40 Ford. Its wonderfully rounded sheet metal and nifty grille work looks great either restored or as a hot rod. Just over 13,000 of these wagons were produced at Ford's Iron Mountain assembly plant. Many of the ones still left have been transformed into stunning rods like this one, which uses a Chevrolet 350 H.O. with tripower, Mustang II front suspension, air conditioning, and a custom Howdy interior. *Owners: Dan and Maureen Wiseman*

maple, or mahogany. The blending of the round, upright but flowing lines of the Fords continued to make them one of the most popular factory wagons.

Packard continued to offer a dual line of station wagons little changed from their introduction in 1937. Buick and Pontiac, along with Plymouth and Chrysler, continued their wagon production into the early part of World War II.

Chevrolet introduced its new wagon also in 1941, promoting it through its corporate ***Friends*** magazine with a "Best Actor" contest. It was typical of glitzy Hollywood promotions of the time. Spencer Tracy won. His prize was a new Special Deluxe eight-passenger wagon. Chevrolet shipped 2,045 wagons that year, all powered by their famous 90hp "Stovebolt Six."

However, World War II was the breaking news, and by late 1941 the auto industry was shifting gears into a wartime production schedule. Huge changes were underway. Ford's Iron Mountain factory was converted to building wooden gliders while, in Detroit, Ford converted to military vehicles, aircraft, and tanks, along with Packard, Cadillac, Oldsmobile, Chevrolet, and Buick, who also converted to aero engines, tanks, and airplane parts.

Some 1942 models made it into production including a new Town and Country edition featuring revised sheet metal and paneling that rolled out to cover the running boards. For two months of 1942 all GM divisions cranked out a limited number of station wagons. Bodies for Chevrolets, Pontiacs, and Oldsmobiles were supplied by Ionia, Cantrell, or Hercules, but by March 1942 all civilian automobile production had ceased.

The government still had a pressing need for vehicles. To help deal with this situation Brooks Stevens designed a conversion using 1942 Ford and Mercury sedans from the government motor pool. Built by

Monart Motors, these Ford wagons were stylish for the time, but more important, they effectively increased the passenger-carrying capacity for these government vehicles quickly and cheaply. Two- and four-door versions were built and were used for regular transportation and as ambulances. How many were built is unknown.

Cantrell and Ford continued building wagons for the military for use as ambulances and radio and reconnaissance vehicles. A special four-wheel-drive Marmon-Harrington Ford wagon was also built for the military through the war. Marmon-Harrington had been building four-wheel-drive conversions since the mid-1930s, so it was natural for them to continue this process with war at hand.

It wasn't just the Big Three in Detroit who offered woody wagons. American Bantam offered a line of small wagons in steel and wood; however, only 322 of these Mifflinburg Body Company-built wagons were sold between 1938 and 1941. Powered by a 22hp four-cylinder engine, the American Bantam never made the impact hoped for, and the company closed up its automobile production at the end of the year. The age of the microcar had to wait until the 1970s, when the Japanese perfected the idea. *Owner: Patricia Retlew*

In 1941 Ford introduced a new body featuring three grille openings, a less peaked hood, and doors that flared out over the running boards. This Super Deluxe Wagon model cost $1,015, making it Ford's most expensive production wagon at the time. It could be supplied with either the V-8 or Ford's new six-cylinder engine. Today, this car is still clad in its original wood but is powered by a 289 OHV V-8 with a C-4 automatic transmission and is used for daily transportation to the owner's construction company. *Owner: Les Layer*

Virtually all wood station wagons in the government vehicle pool found their way into military use. Many were shipped to far off places where they were converted to many purposes.

With the cessation of hostilities there was only a short breathing space before manufacturers returned to building civilian automobiles. A shortage of materials caused immense problems as Detroit resumed production, but wood wasn't one of them, and by the end of 1946, The Big Three all had a mix of wood wagons, sedans, or convertibles rolling off their production lines.

This 1941 Willys 441 Americar Station Wagon is thought to be the original prototype for the three Willys wagons made that year. Built by Mifflinburg, it features maple frames and birch panels with a sculptured mahogany band at the windowsill level. Powered by a four-cylinder 63hp engine and three-speed transmission, this Willys has been fully restored. *Owner: Al Maynard*

As expected, Ford was the first back into the marketplace in late 1945. Ford's vehicles were unchanged from the 1942 models, except for minor trim alterations; however, great things were being prepared on the drawing boards.

Ford had taken notice of Chrysler's 1941-42 Town and Country idea but had a game plan of its own. Henry Ford II took over the running of the company from his grandfather at a most critical time—the company was losing vast amounts of money monthly.

Bob Gregorie, one of Ford's leading designers, had built a roadster woody based on a Model A for young Henry II just before the war. It was this car, and the thought of Chrysler stealing the thunder with the Town and Country line, that prompted Gregorie to design the Ford Sportsman.

The prototype Ford Sportsman was completed in 1945. Henry Ford II immediately stamped his seal of approval on the steel-structured, wood-clad Sportsman design. It was introduced rapidly in 1946 and ran three production years until 1947, with some 1947s eventually being sold as 1948s.

TEXAS 1941
46-400

Left top: The 1941 Plymouth is one of the most elegant and finely detailed production woodys of its time. Its lines were crisp and clean with a fine barred grille opening and ribbing stamped into the fenders. Built by Briggs, the body was constructed using white ash framing over mahogany paneling. Power was supplied via a six-cylinder, 87hp engine. *Owner: Ned Preli*

Left bottom: Buick offered the Estate Wagon on its 49 Special Series chassis in 1941. Powered by an eight-cylinder engine, this sienna rust-colored wagon is beautifully restored to original with an ash-and-mahogany body. Ionia and Hercules supplied bodies to Buick during 1941. *Owner: Bob Brelsford*

The Sportsman had a more important engineering significance than merely competing with Chrysler. It signaled the beginning of the end of wood as a structural component of Ford automobiles. Building products for the war effort had taught the company's engineers new and more efficient ways to design automobiles, and when the Sportsman was prototyped, it was built using a steel substructure and steel paneling, with wood as an external trimming.

A Sportsman version was also offered on a Mercury chassis in 1946, but this only lasted six months. By the end of sales in early 1948 the Sportsman had sold less than 3,750 units; nevertheless, the Sportsman had achieved the desired effect. Not only had Ford sold a few cars, but the Sportsman had raised Ford's image of the better things in life and newer cars to come. Small manufacturers like Studebaker built a couple of prototypes but didn't re-enter the station wagon market. Nash built a line of Suburban "wooded" sedans, while Willys and Crosley dabbled with the idea and eventually both built steel-bodied wagons with a wood-look structure using Di-Noc paneling.

Chrysler re-entered the marketplace with a mass of Town and Country advertising. This line was trumpeted and advertised as a two-door hardtop, a two-seat roadster, a four-door sedan, a convertible, and a Brougham. However, only the convertible and the four-door sedan made the production line in 1946.

Built on both Windsor and New Yorker chassis, the various Town and Country models featured Chrysler's huge straight-8 engine. Framed in white ash with mahogany paneling, they were the delight of the social set.

Chrysler offered the most innovative selection of woodys. The introduction of the 1941 Town and Country signaled a new age of style for the American automobile. This 1942 model featured revised sheet metal and trim from the 1941 model. Just 1,000 were built in a mix of six- and nine-passenger models. *Owner: P. Montano*

The combination of timber looked very smart and was tightly installed because of the enormous amount of hand labor required for every vehicle. In an attempt to lower the amount of hand-work, engineers invented a process that adhered the mahogany veneer to a metal panel before it was formed in a press.

The most popular of all Town and Country models turned out to be the convertible. By 1948, a surprising 8,368 had been built of a vehicle that cost close to $3,000

Hudson had been selling woody wagons for four years by 1942, when this vehicle was sold. Based on the most expensive four-door Series 21 Super Six in the Hudson line, this wagon was built for Hudson by Campbell Body in Waterloo, New York. It has been fully restored and is one of only two known to exist. It is powered by an L-headed six-cylinder, 102hp engine. *Owners: Jim and Betty Fritts*

when a Ford Sportsman could be had for $2,285. The Ford Sportsman and the Chrysler Town and Country series were aimed at an even higher-end market than the expensive wood wagons built before World War II, but they had the desired effect of raising the company's image.

They were "the kind of cars dreams were made of" with their up-market look offering instant luxury and style in a marketplace suffering from outdated pre-war styling and postwar shortages.

Plymouth plodded on with its prewar wagon into 1947 while Dodge stayed clear of the wagon market until

The Town and Country was a revolutionary concept using a steel roof, four doors, and a barrel-back trunk enclosed by clamshell wooden doors. The original concept had come from Paul Haufer at the Boyerstown Body Works and was carried into production by David Wallace, president of Chrysler. *Owner: P. Montano*

1949. However, a buyer could get a Dodge wood wagon built privately from an outside supplier like Cantrell on a light truck chassis.

At GM the story was quite different. Even though the production numbers were small, all the divisions offered station wagons. Chevrolet was back with a Fleetmaster-based wagon with bodies from Ionia, Fisher, and Cantrell. At the same time, Pontiac and Oldsmobile offered Ionia and Hercules bodies while Buick sold its Roadmaster 79 Series Estate Wagon fitted with a special Ionia body.

The 1946 Pontiac Streamliner Eight Deluxe station wagon was the most dynamic looking of all GM's woody wagons. It was styled with rear fender skirts, white ash framing, mahogany paneling, fine chrome trim, and under the long hood hid a 103hp straight-8.

By the late forties automobile styling started to catch up with the new postwar engineering. Huge changes were coming, and as sheet metal moved up the bodywork, less emphasis was placed on wood panels and structure.

In 1946, Nash followed Chrysler's style with a sedan woody featuring wood paneling that imitated the original "barrel-back" Town and Country. During its three years of production, only one thousand of these Nash Ambassador Suburbans were built, but it once again showed how significant the woody theme was to the manufacturers.

The aftermarket also came into the act, offering Sportsman-style wood cladding kits for a variety of non-factory woody models. One of the best known was the Country Club Kit, which turned a 1947-48 Chevrolet Fleetmaster convertible into a Sportsman look-alike for $295. Another similar kit was available for the Chevrolet Aero sedan.

These kits were sold through Chevrolet dealers and apparently were designed by the Fisher Body division, but they were manufactured and sold by an outside supplier, Engineered Enterprises in Detroit. The Bellbod Company in New York also sold aftermarket "woody-izing" kits for a variety of forties vehicles. These kits consisted of hardwood framing and mahogany paneling, both of which were simply screwed to the side of the vehicle.

Ford continued on with the "fat-fendered" look, revising much of the trim and crisping up the lines of the wagons. Different wood paneling combinations were offered, including Philippine ribbon mahogany and

With the start of the war, Ford's production of new 1942 models was curtailed abruptly when production of civilian automobiles ceased in March. This made models like the Super Deluxe extremely rare, but their interesting new front-end styling and good-looking body work made them even more popular. Only 1,222 wagons were built.
Owner: Dick Grace

maple framing. By 1946 both Ford and Mercury had their own versions.

In 1947, the Deluxe version and the Super Deluxe Fords differentiated with the Deluxe using lighter-colored exterior birch and the Super Deluxe using a mahogany-look paneling. The dash paneling was trimmed in a maple-grained finish while the roof was still slatted with white ash ribbing.

Both models still featured soft roof panels with padded tops over inner wood frames. These wagons were built by the 300 or more craftsmen at Iron Mountain.

The woody market was not just all regular wagon-style vehicles. GMC, Chevrolet, and International Harvester offered truck-based woody buses and full-sized carry-all suburbans with bodies built by Cantrell and Ionia.

By 1948 the automobile industry was once again approaching full speed. Packard introduced one of its most stylish wagons with the Station Sedan. This steel structured four-door, six-passenger wagon was built by Briggs manufacturing and featured a chunky set of ribbed door panels topped by beautifully styled side window frames that flowed into a wood-framed tailgate.

Designed by Al Prance, this gracious model won many international design awards, including a gold medal from the New York Fashion Academy for the interior, and styling awards from shows in Monte Carlo, Italy, Switzerland, and Venezuela. Unfortunately, this

The fat-fendered look became the hot rod rage in the mid-eighties, and this stunning black 1946 Ford Station Wagon is a perfect example of that concept. Powered by a small block Chevrolet engine and rolling on an independent GM front suspension, this hot-rodded 1946 offers air conditioning inside its eastern hardrock maple-framed body, which is paneled in ribbon-striped African mahogany.
Owner: Jim DeFrank

More OF EVERYTHING YOU WANT WITH *Mercury*

For more real fun, here's the car for you—the spirited Mercury Station Wagon. You'll like its country club styling . . . and its practicality too! Here is the answer to all your motoring needs. More capacity . . . it carries eight in comfort! More usefulness . . . it does utility hauling! More beauty . . . its youthful smartness is equally at home in town or country! Drive it, and see why Mercury gives you more of everything!

MERCURY—DIVISION OF FORD MOTOR COMPANY

MORE ROOM Look inside! Ample room for eight, with space to spare! Those comfortable seats are genuine leather, skillfully tailored in a choice of colors: tan, red or gray. It's a big, handsome car, for big families with lots of friends!

MORE STYLE Superb design and coachwork give the Mercury Station Wagon a smart, custom-built look. Maple body framing is combined with birch or deep-toned mahogany panels, for extra richness. A stand-out, wherever it goes!

MORE CONVENIENCE By far, the most useful type of car on the road! Center, rear seats are removable to triple storage space. Sloping tail-gate lowers for easy loading and added length. Here's all the room you need, and more!

TUNE IN . . . THE FORD–BOB CROSBY SHOW—CBS, WEDNESDAYS, 9:30-10 P.M., E.S.T. . . . THE FORD SUNDAY EVENING HOUR—ABC, SUNDAYS, 8-9 P.M., E.S.T.

This 1946 Mercury ad shows woody owners reveling in the good life.

In 1946 Pontiac made the station wagon part of its Streamliner Series 28. Bodies were built by Ionia and were among the first to take up the chrome decoration theme again after the war. This Catalina Cream wagon features its original but restored ash-and-mahogany body and is still powered by the stock 103hp eight-cylinder engine. *Owners: John and Demere Bates*

was about as good as it got as the Station Sedan was expensive and sold in low numbers. It was also the final year of the wood-framed Plymouth wagon, which had been reissued with only minor changes since 1946.

All the wooded station wagons had structural and maintenance problems, but as the amount of wood diminished, so did the problems. In wet weather the wood swelled, and in hot weather it shrank and squeaked. Detroit still had not resolved the need to refinish the wood once a year to maintain its luster.

In some ways the demise of the woody was a three-way street. The postwar population no longer wanted vehicles needing special servicing, Detroit wanted out of the "furniture" business, yet they both still wanted the styling.

To this end Oldsmobile wrapped up its last true wood station wagon and replaced it in 1949 with a completely new steel-bodied Futuramic 78 series designed by Harley Earl. It could be optioned with the now famous 303ci Oldsmobile Rocket OHV, which developed 135hp. The new Oldsmobile featured a touch of wood trim around the belt line and tailgate.

Nash came into the Town and Country look with its 1947 Ambassador Suburban. Nash was looking to upgrade its image, and the Surburban was a cost-effective way of creating a new model without having to build expensive steel tooling. The Ambassador Suburban featured ash framing and mahogany paneling, which imitated the style of the original 1941 Chrysler Town and Country. The Suburban was an expensive model, like the Town and Country, and only 1,000 units were sold during its 1946 to 1948 production years. This strato blue version is powered by a six-cylinder engine. *Photo and owner: James Dworschack*

A 1947 Pontiac ad.

Buick and Pontiac continued to build fairly conventional woody-style wagons, leaving Buick as GM's last vestiges of the horse-drawn era.

In late 1949 Chevrolet introduced a new line of wagons to replace their "wood" wagon. This new all-steel wagon featured "wood-grained" body moldings that simulated the traditional wood framing, just as Oldsmobile had produced with the Futuramic.

Chrysler was slower to follow. The Chrysler Division, DeSoto, and Dodge reintroduced the station wagon model in 1949 to join Plymouth with its dual line of all-steel wagons and separate Town and Country mod-

Left: The 1946 Ford Station Wagon has become the favorite hot rod woody. This gorgeous example was built at Roy Brizio Street Rods in South San Francisco for Nancy Edelbrock, wife of speed equipment manufacturer Vic Edelbrock. Its nostalgic wide-whites and ruby red paint give it a fifties period look while its small block offers plenty of horsepower. *Owner: Nancy Edelbrock*

Left bottom: The Town and Country convertible is one of the most collectable of all wood-sided vehicles, especially as a convertible. Based on the New Yorker chassis, the Town and Country was the vehicle to be seen in during the late 1940s. Powered by an eight-cylinder flathead engine the T&C was no powerhouse, but this didn't stop it from being a favorite among the rich and famous. *Owner: Robert Brelsford*

Bottom: The Royal Maroon paint on this 1947 Town and Country four-door contrasts beautifully against the wood on this sedan, which features every option Chrysler offered for the model, including reversing light, roof rack spotlight, radio, and sun visor. *Owner: Lloyd F. Mayes*

Left: This stock 1947 Ford Super Deluxe station wagon features a Ford maple-framed, mahogany-paneled body and is powered by the factory 239ci V-8 under the hood. Offered in two models, the Deluxe and Super Deluxe, they were built at Ford's Iron Mountain facility. The 1947 Mercury shared the same body but with a different front end treatment. *Owner: David Doherty*

Gene Autry was one of Hollywood's great western screen heroes. His love of the West, horses, and riding spread throughout his life. This was his personal 1947 Mercury station wagon, which he used to tow a custom-made 1940 horse trailer. Gene is on the left with his famous horse, Champion. This photo was taken around 1949. *Photo: Gene Autry Western Heritage Museum*

els. These wagons featured an unusual molded-steel spare tire carrier contoured into the tailgate. The Chrysler Royal station wagon looked like and was a logical extension of the Town and Country theme with its heavy-set ash framing and Di-Noc inserts.

The Town and Country was given a facelift in 1949 when Chrysler introduced its first complete redesign after the war. The new body featured a flatter hood with blended bodywork. It was a far less attractive automobile than the earlier version, but it brought with it a mass of engineering improvements.

Its white ash framing was now purely ornamental, and the wood paneling had been replaced with the new Di-Noc wood-grained adhesive plastic. The first 700 production units of the Town and Country were built this way; but the Di-Noc paneling was soon dropped, and the sheet metal was painted in a body color that produced a look similar to the earlier Ford Sportsman.

The Town and Country wood framing was assembled before being fitted to the body, which required extensive hand-formed contouring of the compound curved frames so they mated to the metal body parts correctly. The cost of hand-building each body became too expensive, and bigger profits could be made more easily from all-steel automobiles.

Several other important factors were also stacked against the continuance of the Town and Country. One was the overwhelming cost of replacement parts. Most of the body pieces ran about five times the cost of conventional body panels, and these stylish monsters of the social set were also weighed down by primitive flathead power and a four-speed, semiautomatic Fluid

Before World War II, Marmon-Herrington built four-wheel-drive conversions for a number of years, based on Ford sedans and wagons, for use by timber companies, the National Park Service, and some early East Coast ski resorts. During the war some were built for the military. This 1946 Mercury was built for naturalist Don Bleitz to use on his Sierra Mountains research trips. It was set up by Coachcraft of Hollywood as a camping wagon with a folding bed, ice chest, water tank, storage cupboards, roof rack, and roof access step. The maple and mahogany trimmed wagon was recently restored by the owner to its original setup. *Photo by owner: Dave Holmes*

Drive, which needed coaxing to shift smoothly and on time.

Weighing in at well over 4,000lb, and with some models closer to 5,000lb, they offered only a 114hp 6 or a 135hp 8, and thus delivered a lackluster performance from their 35-to-1 power-to-weight ratio. (A comparable luxury Chrysler LH sedan today offers a snappy 15-to-1 power-to-weight ratio.)

Meanwhile, Ford was about to set the wagon world on fire again in 1949 with its first all-new postwar station wagon.

Ford managed to delete some of the flaws that had bugged traditional wood wagons. The new Ford/Mercury wagons featured an all-steel structure that practically stopped the body squeaking and much of the wind noise, and it theoretically extended the life of the wagon by designing the body with bolt-on, replaceable wood panels.

Until this stage all Ford station wagons had used solid maple framing, but the 1949's framing was steel covered with paneling created using the latest in electronic technology. These were formed using a microwave bonding press that squeezed a group of phenolic resin-coated wood pieces together to form a frame blank. The panels were created with a similar process that used an outer layer of maple over an inner layer of ash. This allowed the panel to be formed easily.

These panels were cured for only a short time and then trimmed and shaped to size. Ford and Mercury shared bodies, but the Mercury featured different front sheet metal, front doors, and interior trim.

The concept was actually a great idea, but it failed to make the grade. The paneling deteriorated like the earlier versions and still needed an annual or biannual varnishing to maintain its luster. Those who didn't keep up with its required maintenance ran into other problems. Ford hadn't kept sufficient replacement panels on hand when they were really needed four or five years later, and it was from this point on that many of these wagons started down the hill to the junkyard or as transportation for the surf generation.

"WE'RE ONLY GOING TO BE GONE OVER NIGHT."

An ad for the 1946 Ford Sportsman.

NEW 1946 FORD SPORTSMAN'S CONVERTIBLE!
Outside and inside, there never was a car like this before! The new Ford Sportsman's Convertible is really two cars in one! Ford designers have combined the paneled smartness of the station wagon and the touch-a-button convenience of the convertible!
JUST TOUCH A BUTTON
. . . and in 30 seconds you have a snug "sedan" that's weather-tight! With the top up or with it down, the Sportsman's Convertible is a honey for looks!
THERE'S A Ford IN YOUR FUTURE
NEW OVERSIZED BRAKES
. . . self-centering and hydraulic . . . with big 12-inch brake drums . . . to give you smooth, straight-line stops on all road surfaces. Quiet! Easy-acting. Long-life linings!
UNE IN . . . The FORD–Bob Crosby Show — CBS, Wednesdays, 9:30-10 P.M., E.S.T. . . . The FORD Sunday Evening Hour — ABC, Sundays, 8-9 P.M., E.S.T.

The Ford Sportsman first arrived on the market in 1946. Henry Ford II was now in charge of the company, and he liked the idea of being able to offer a Town and Country-type convertible without having to build extensive tooling to create a new model. All the timber work was decorative using maple framing and mahogany paneling. Only 3,525 were built during its three years of production. The Sportsman is a highly valued collectors' car today, and this 1947 example belongs to the acclaimed Towe Ford Museum near Old Town Sacramento, California.

Right top: Oldsmobile was also back in the station wagon business after the war. This Special 66, one of 968 wagons made in 1947, was purchased new by the father of the current owner from the Watertown, Massachusetts, Oldsmobile dealer for $2,500. The all-birch Hercules body is stained in walnut and features a three-seat interior. *Owner: Alfred L. Maurer*

Right bottom: Ford and Chrysler offered sporty "wooded" convertibles in the mid-forties, and the aftermarket also got into the game with the Country Club, a dealer-installed wood package. This 1947 Chevrolet Fleetline Country Club convertible features a reproduction of the $295 kit. Companies, such as Bellbod in Brooklyn, New York, and Engineered Enterprises in Detroit, sold kits for a variety of models. *Owner: James Ashworth*

Massachusetts
OLDS·47
The Spirit of America

STOCK 47

Left top: The 1948 Ford wagon varied little from the 1947 model. A matching Mercury version was offered, and it was these woodys that gave the surf scene its classic transportation. Even today Earl Hewes uses his restored wagon for surf cruising, with the longboard sticking out of his wagon. Earl won the Old-timers Longboard Championships in 1992 at Huntington Beach, California. *Owner: Earl Hewes*

Left bottom: The popularity of 1948 Ford wagons extends far and wide. This snappy red wagon from Spokane, Washington, was found via a classified ad. It was complete and drivable, and today it's restored and rodded with a 327 Chevrolet V-8, dropped axle, and good hot rod upgrades. Most of the wood is original and shines against the Chevrolet rally red paintwork. *Photo: Peter Vincent. Owner: Len Bush*

The Town and Country offered about as much elegance as Detroit could muster in the postwar years of the forties. The 1948 convertible model was unchanged from the 1947, but a few changes were made to the sedan. The wooden parts were built by Pekin Wood Products in Helena, Arkansas, and then shipped to Chrysler's Jefferson Avenue plant in Detroit for assembly. Powered by the Spitfire eight-cylinder and Fluid Drive transmission, the T&C offered plenty of lugging power but little in the way of pep. *Owner: E. Hawbaker*

Leo Carrillo, the famous sidekick of " the Cisco Kid," owned this steer-horned 1948 Town and Country convertible. The horned hood was used for parades and appearances. It is seen here in Harrah's Museum in Reno, Nevada, where it was on show for years.

Right top: The Chevrolet Fleetmaster station wagon must have been a popular model in 1948, as it sold 10,171 units for just over $2,000 each. It was an upscale model with its long flowing bodywork built in ash with mahogany paneling by both Ionia and Cantrell. The ceiling was ribbed wood and the top was covered with a leatherette material.
Owner: Bob Weaver

Right bottom: This 1948 Fleetmaster station wagon was converted into an ambulance for use by the U.S. Bureau of Mines in Rifle, Colorado, in the late 1940s. It then saw service with the Grand Valley Volunteer Fire Department before the current owner purchased it. It is still fitted with its working hardware, including lights, siren, and stretcher.
Owner: David Doyle

AMBULANCE
GRAND VALLEY
Volunteer Fire Dept.
AMBULANCE

Chevrolet had new plans for steel station wagon models, and 1948 was the last of the all-wood-bodied Chevrolet wagons. This beautifully restored wagon features a rare Fisher body in ash and mahogany with three rows of seats, split rear bumper, and a tailgate-mounted spare. *Owners: Jim and Dorothy Bartish*

Packards had always offered style and power, but at a price. In 1948 the company offered its first new postwar model wagon called the standard Eight Station Sedan. The wonderfully styled tailgate had a deco look as it rolled around the roof and formed into the new woodwork. Fewer than four thousand were sold during its two-year production run. *Owner: Jeremy Janss*

The 1948 Packard Station Sedan turned out to be one of the most gracious-looking wagons of the period, predating Ford's new look by a year. The four-door six-passenger wagon was spun off the production Packard sedan by changing the roof panel and trunk lid and then creating the wagon conversion out of ash and maple. Designed by Al Prance and built by Briggs Manufacturing for Packard, it was powered by a 130hp L-headed straight 8. The Packard Station Sedan sold only to the well-heeled social set with a price tag of $3,424, and less than 3,900 were sold during its three years of production. *Owner: Joe Scott*

The 1948 Plymouth station wagon had been carried over virtually unchanged since its introduction in 1946. Hercules was the body supplier, and this special deluxe model was equipped with a reliable 95hp Spitfire six-cylinder engine. *Owner: Joe Flores*

Right: A 1949 Ford advertisement.

New 'Mid Ship' Ride
The finest, the costliest cars in America seat you in the smooth riding center-section. So does the new Ford—with a new "Mid Ship" Ride!
New 'Lifeguard' Body
The finest, the costliest cars in America have all-steel bodies with multiple member frames. So does the new Ford! Its new "Lifeguard" Body in combination with the 5-cross-member box-section frame is 59% more rigid.
LIFE GUARD
The '49 Ford's out Front
...the fine car of its field!
Up to 10% More Gas Savings
The finest, the costliest cars in America are powered with V-8 engines. Ford offers your choice of V-8 or Six. Stepped-up economy, too—up to 10%.
New 'Hydra-Coil' Springs
The finest, the costliest cars in America have independent front wheel spring suspension. So does the new Ford. New "Hydra-Coil" Springs in front, new "Para-Flex" Springs in the rear.
New 'Picture Window' Visibility
The finest, the costliest cars in America accent visibility. So does the new Ford with "Picture Windows" all around. Bigger windshield with narrower corner posts plus a rear window that's windshield-big itself!
There's a New Ford in your future
FORD
White side wall tires, optional at extra cost.

MERCURY
CALIFORNIA
WOULDE
NATIONAL WOODIE CLUB

Left top: Ford's new wagon in 1949 was its first completely new postwar midrange model. It featured an all-steel roof and was only available in a two-door. This hot-rodded version runs a "built" 1951 Mercury flathead V-8 and a C-4 Ford automatic transmission. The top has been chopped 2.1/2in and is framed in eastern hardrock maple with African mahogany paneling and bird's-eye maple interior. *Owner: Corbin Taylor*

Left bottom: In 1949 Mercury shared the same Iron Mountain-built body with the 1949 Ford; however, the front doors weren't the same for the Mercury due to the front fender design, which flowed differently into the door panel. This 1949 Mercury was used by Barry Goldwater as his campaign wagon. Goldwater had a special roof rack installed to mount a public address system so he could stand on the roof to give speeches. The wagon is very original and features the optional sun visor, spotlight, and grille guard. *Owner: William Jacobs*

Mercury offered only a two-door in 1949, and, interestingly, the taillights on this model featured a special articulated lever mount that kept them horizontal as the tailgate was lowered, allowing the lights to be seen if the wagon was driven with the tailgate down. This original wagon is still being used as a "surfer's woody"—the current (fourth) owner has surfed along the California coast for the past thirty years. *Owner: Don Iglesias*

You know, I think I liked it better in the crate.

Left: This gorgeous 1949 DeSoto Deluxe is one of 680 made that year. Fully restored and trimmed in factory deep red, it is powered by a six-cylinder flathead Chrysler, developing 112hp. It was fitted with the $50 optional third seat, which made this DeSoto into a nine-passenger. The spare tire was mounted in a case that formed part of the all-steel tailgate. *Owner: Thomas H. Doyle*

The 1949 Plymouth station wagon was basically the same vehicle as the DeSoto that year, with the exception of the tailgate and the rear bumper. Bodied in ash and mahogany, these rather formal-looking Chryslers were a perfect wagon partner to the Town and Country. *Owner: Barry Foster*

The Town and Country had reached its peak in 1947, and by 1949, production had declined to 1,000 units. Only the first 300 were finished in Di-Noc mahogany paneling. This Ensign Blue convertible was built late in the model run and features painted body paneling, which gave the line a new look and a new life. It is fitted with a padded dash, Kelsey-Hayes wire wheels, Vacu-Ease power brakes, dual mirrors, heater radio, and back-up light. *Owner: Thomas White*

Right: Chrysler went to France to create the ad for their late 1949 Town and Country.

Beautiful

The New Chrysler

LONG . . . LOW . . . AND LOVELY

The most spacious and gracious of all convertibles. Introducing new riding luxury you can't believe until you feel it! All steel for the first time—except for stunning trim of polished white ash—the new Town & Country is the most beautifully engineered Chrysler yet! With the simplest automatic drive of all... with the first completely waterproof ignition system... with over 50 big basic, exciting advances! You'll marvel at the 15-second operation of its top! You'll find a rainbow of thrilling colors available with swank leathers to match or contrast. Phone your nearby Chrysler dealer today.

CHRYSLER DIVISION • CHRYSLER CORP. • DETROIT

Town & Country

This very rare 1949 Buick Super Estate Wagon was a new model for Buick with its all-steel roof, wood framing, windows, and tailgate. The ash-and-mahogany bodies were built and installed by Ionia. Less than 1,900 were delivered. It is powered by a Buick straight-eight and a Dynaflow transmission, which is needed to carry its 4,100lb. *Owner: Jim Grace*

Right: This beautiful custom-bodied Series Seventy-Five, seven-passenger 1949 Cadillac is the ultimate in forties California custom coachwork. The Cadillac was one of about six such Cadillac woody limousines built by Maurice Schwartz in Los Angeles for Metro-Goldwyn-Mayer (MGM) movie studios. Schwartz, a former partner in the famous Bohman and Schwartz coachbuilders, opened his own shop after his partner retired in 1947. These Cadillacs were used to tránsport movie stars and executives to and from movie sets, railway stations, and airports. This Cadillac was involved in a roll-over accident during a trip to a movie location at Big Bear Lake in about 1954 but was not repaired for another thirty years. In the early eighties it was recovered from a field and completely restored in Sacramento. *Owner: Ramshead Collection*

This rear view of the last of the real woodys shows how beautifully styled the 1953 Buick Super Estate was from the rear. The Skylark Kelsey-Hayes wire wheels and wide-whites, along with the contrasting green bodywork and white ash woodwork, make a startling combination. *Owner: Kent Berge*

Chapter Five

The Fifties

The beginning of the 1950s trumpeted a new world order. The Korean War was just over the horizon, and the economy was picking up speed at a rapid rate, but Detroit still was having trouble filling its postwar orders.

The Town and Country made its last debut in 1950. Chrysler offered a single Newport two-door hardtop Town and Country on a 131in wheelbase powered by an eight-cylinder engine. Only 698 were built and sold, which was not surprising considering its $4,000 price tag when a conventional Chrysler Royal Sedan cost $2,136.

DeSoto also had a line of wood station wagons. Two versions were offered; one for domestic consumption, known as the Custom, and the Diplomat, which was for export. Both versions were based on the Plymouth Royal. The most intriguing feature of the Royal series was the first production appearance of the "crank-down" disappearing tailgate window. These wagons were framed in white ash with mahogany Di-Noc paneling.

Chrysler continued its six-passenger Royal station wagon and sold only 599. By the end of the model year the Town and Country and the Royal wagon were headed to the woodpile, too, sunk like scuttled land yachts.

Buick's station wagons came complete with "portholes," but the body still featured real wood. The company's Roadmaster and Deluxe used only a small amount of side-wood trim; however, the tailgate could still be considered a wood product.

Chevrolet, Oldsmobile, and Pontiac had already crossed the boundary from wood to steel by 1950 using Fisher-built wagon bodies. These bodies continued the woody theme with wood frames pressed into the metal bodywork. The top-of-the-line Styleline DeLuxe station wagon was the only model that featured simulated wood graining on these frames. The wood graining was applied using a photo transfer process. Oldsmobile and Pontiac had similar offerings, which continued unchanged until the 1955 model year.

Woody station wagons were not just an American product. The British and French both had their own variations on the same theme through the same years. The French built some interesting Simca woody wagons based on Fiat Topolinos along with some Delahayes and Citroëns.

Across the channel the British were also sawing away, building custom shooting brakes and estate wag-

Woodys were really befriended by surfers, as this photo from the summer of 1953 shows. One of those was Doug Craig, an early California surfboard builder; he's on the far right with his buddies John Hall, Stu Skeele, and Don Guild. They would load up their Craig balsa boards in Hermosa Beach and head out south looking for surf. Doug recalls: "We used to pile a camp stove, some grub, surf trunks, and boards into that old 1950 Ford, so we were equipped to camp on the beach and many times would end up at San Onofre for the weekend." Photo: *Bobbie Craig; Tom Craig Photo Library*

ons for the lords and masters of the great estates to use as hunting and fishing wagons. They used a wide variety of marques including Ford Pilot V-8s, Rolls Royces, and Bentleys.

The Queen Mother had a Ford Pilot shooting brake built especially for her to use on the royal family's estates in England and Scotland.

In the United States the new Ford station wagons continued to sell quite well for just about $2,000. In 1949, Ford sold just over 30,000 units; in 1950 a tad over 20,000; and in 1951 the number jumped up to nearly 30,000 again. Ford began trimming down the wood content of the new model, and, by the end of the 1950 model year, the wood tailgates were replaced by a steel panel trimmed with a wood-grained decal.

The interiors of the Fords also lost some of their wood magic. The wooden door panels were replaced by imitation leather, and the wood-grained dash was eliminated along with the steel-cased spare tire carrier.

At the end of 1950 Ford closed its Iron Mountain facility, and for the first time in twenty years, Ford station wagon bodies were manufactured by an outside supplier, Ionia Manufacturing Company in Ypsilanti, Michigan.

Yet, the market was mutating and the public wanted more from Ford. For a start, the consumer demanded four-door wagons that did not have maintenance problems. GM and Chrysler were selling far more wagons than Ford, and only one of them was a traditional woody model.

In 1952, when Ford introduced its all-new Gordon Buehrig-designed line-up, it joined GM's line of thought using wood as a light trim option for the new wagons. Both three- and five-door models were introduced, although Mercury offered only a five-door version. The highline Ford Country Squire and Mercury Custom Series wagons were both trimmed in wood.

Detroit no longer liked wood. The new steel age had arrived. There was plenty of it, and it didn't rot, fade or peel, but outside suppliers were still very active. Builders like Cantrell offered a collection of standard model bod-

In 1950, Ford made only a few minor grille and interior changes to the StationWagon. It was still powered by the 100hp flathead V-8 and had only two doors. This beauty belongs to the owner of Heiden's Woodworking in Encinitas, California, and is all stock with original wood but now powered by a 350 Chevrolet. Note the custom surfboard-shaped interior rear-view mirror. *Owner: Ron Heiden*

ies for all kinds of commercial chassis including One Ton GMCs, Chevrolets, and Dodges. Cantrell and Huntington had become the last vestiges of the carriage-building trade where craftsmen hand-built automobile bodies.

Buick's Roadmaster and Super Estate wagons were now the last GM wagons to use traditional woody framing in white ash with mahogany paneling. These wagons were built at Ionia, using a partly assembled Buick body that was shipped to Ionia, built up as a station wagon, and then shipped back to Buick complete, ready for final assembly. In all, Buick sold 2,500 wagons that year, and these were the last of the "wooden dinosaurs" built by or for any Detroit manufacturer.

Chevrolet, Pontiac, and Oldsmobile shared the main body section and sheet metal of their wagons until 1953. In 1954, Chevy and Olds introduced new models

Built on a Chrysler Windsor chassis, the 1950 Royal station wagon was an elegant partner to the last of the Town and Countries . It features a crank-down rear window in the tailgate, steel roof, ash framing, and Di-Noc mahogany-paneled doors. *Owner: Jim Grace*

with their own style, while Pontiac carried on for another year before introducing the new wagon models.

Each of these pre-1955 wagon models offered a touch of wood with a section of photo-stencil wood graining around the belt line of the upscale models. From 1955 it was hands-off at GM for any wood-look trim. Chrysler also dropped the wood look and went into fins, chrome, and multicolor paintwork for its most decorative models.

It was now up to Ford to carry the banner. The company had created and held the woody market in its hands for over twenty years, and now, with the advent of so many new materials and pressing consumer needs for family automobiles, it pursued the station wagon as a new status symbol.

Henry Ford II still had quite a fancy for the woody look and oversaw the continuation of the theme as a Ford styling cue. Ford continued to use real wood trim to surround the artificial wood paneling on both the 1954 Ford and Mercury models.

These new station wagons carried the Country Squire tag while the Mercury version was tagged the Monterey. The Monterey was sold as a five-door only, in either six- or eight-passenger seating configurations. These models featured woody styling, which was rather heavy handed considering the more refined look of the GM products.

Several British manufacturers also offered woodys. Austin and Morris, as well as the Standard Motor Company, turned out small woody wagons during the fifties in England, including Morris with its Traveller; Austin with its A-Series; and Standard with its tiny 8 and 10 models.

Allard, a small manufacturer of race cars and touring cars, produced the P2 Safari wagon in 1953. These gracious three-door, four-seater wagons were built on a tubular chassis and were powered by a variety of engines, including the flathead Ford V-8. Apparently less than ten were built, but their style rates as one of the most gracious woody designs of all times.

In Detroit the switch to "fiberglass wood" framing on the Fords in 1955 was double-headed. It meant that servicing the wood on a regular basis was unnecessary, and it also allowed the design staff the opportunity to create a more refined look to the framing. They could now make it any size or shape. The problems of warping or cracking because the wood was too thin or could not be made in a desired shape became obsolete.

This was the first year of the new look, "simulated boat-deck planking" wood paneling. The combination was actually quite startling on either a white or a bright red wagon because it was also two-toned in a complementary color around the windows. The Mercury version looked completely different to the Ford version as it featured double panels set in two levels, fiberglass frames, and simulated mahogany wood paneling.

As the 1950s continued, Ford evolved the Country Squire image in both lines with more refinement and some interesting changes in wood grains and framing styles. In 1957, Mercury took up the Colony Park tag for its upline woody wagons, which featured pillarless, hardtop styling and a very interesting reversed-cove fin design.

The introduction of the Edsel in 1958 brought with it the top-of-the-line four-door Bermuda nine-passenger wagon which featured two-tone paint. It was one of the most interesting "fake" woody styles that ever mutated out of Detroit it with its combination of multiple colors, fins, and wood trimming.

Ford continued to be the only manufacturer using wood trimming with the new 1959 wagon featuring a huge section of simulated wood paneling surrounded by a heavy fiberglass fake ash frame.

The British public had also discovered the joys of motoring wagons. While the custom-built shooting brakes were part of the world of the wealthy, the station wagon was more the means of transportation you might find a farmer using. Austin of England had produced a series of sedans and wagons commencing after the end of the war. By the beginning of the fifties, they were delivering the A70 Hampshire Countryman, which had evolved from the A40 Somerset wagon. The A70 featured oak framing with dark Masonite paneling and suicide rear doors. *Photo: British Motor Industry Heritage Trust*

Previous pages: This 1951 Chevrolet four-door station wagon is a stunning example of the extremes to which rodders will go to build perfect machines. It started out with a 6in chop to the roof and doors and ended with shaved body work, dechroming, 455 Oldsmobile power, a Mustang II front end, and custom interior. It was finished off in gloss black and then highlighted with some of the finest "wood graining" ever seen on an automobile. The graining runs around every door jamb and was done to completely simulate what the body would look like if it had been done in timber. *Owner: Roger Ward*

Right top: Chevrolet shared its basic wagon body with Pontiac and Oldsmobile. These all-steel wagons were built by Fisher Body using sheet metal that featured stamped framing. They were then trimmed using simulated wood-graining decals. This 1951 Styleline Deluxe is finished in Aztec tan with wonderfully detailed ash and mahogany wood graining. *Owner: Jack E. DeLuca*

Once again, in 1951, Ford shared its wagon body with Mercury. It features a new grille treatment with twin-spinners. These wagons were now being built by Ionia as Ford's Iron Mountain plant had closed at the end of 1950. This Country Squire is "surfer-ized" with flames and is used to promote Woody's Diner in Huntington and Sunset Beach. *Owner: Mike Chase*

Right bottom: This 1949 Ford station wagon was fitted out as a photographer's car for Quinn Publications of Hollywood, publishers of Hop Up, Motor Life, and Rod and Custom magazines. It had been lightly customized by Barris Kustoms and featured a photographer's platform, accessible using a pair of steps up the rear C pillar. Seen here at Bonneville Salt Flats for Speed Trials in the early '50s, the Ford later appeared on the cover of the March 1955 Rod and Custom magazine, decorated with bright red flames running over the nose. *Photo: Dean Moon*

13405

HOP UP & Motor Life Magazine
Rod & Custom
QUINN PUBLICATIONS Inc.
HOLLYWOOD

Only Pontiac and Chevrolet shared bodies in 1954 for the wagon. The Pontiac Chieftain Deluxe featured wood-grained white ash framing with a fine mahogany-grained insert. This wagon model could be ordered as a six- or nine-passenger model. 1954 was a big year for Pontiac: the 5,000,000th automobile was built. *Owner: William R. McMillen*

Right bottom: This 1953 Buick station wagon was the last of the real woody wagons from General Motors, or any other major American manufacturer for that matter. Built by Ionia, its gracious proportions and fine woodworking did not stop its decline. This Super Estate Wagon is one of 1,830 made, is finished in Mandarin Red, and features an ash-framed body built and installed by Ionia. It is powered by a 322ci V-8 and is fitted with a Dynaflow transmission, power steering, and brakes. *Owner: Allan Dann*

Right top: From the twenties, Morris and Austin had been heated rivals in the British motor industry; however, in 1952, under the leadership of Leonard Lord, the companies merged to form the British Motor Corporation, headquartered in Longbridge. This provided the opportunity to develop new and interesting products, including the Morris Isis series. This Series 1 Isis Traveller is shown in a very British rural scene with the farmer chatting to his farm manager as his workers stack hay. Like the 1950 Chrysler Town and Country, the Traveller only used decorative oak framing over painted sheet metal. The Isis was built using cab and chassis construction, which allowed Morris to easily build either a pickup, van, or wagon on the same chassis/cab unit. *Photo: British Motor Industry Heritage Trust*

XWL 58
GRX911

The Allard P2 Safari is, without doubt, the most gracious wooded wagon anyone had ever built. Allard was a small British manufacturer who had won some amazing races against the world's best, including the Monte Carlo Rallye in 1952. In the early to mid-fifties, Allard delivered a small run of handcrafted P1 touring cars and P2 Safari models based on their racing car chassis designs. Allards were all handmade, and this stunning 1954 P2 Safari is one of eleven made during its production years. Powered by a Ford "Koln" 3.9 liter V-8, the Safari's flowing aluminum bodywork rolls back into a wonderful blend of English ash framing and mahogany paneling. Photo courtesy of Thoroughbred and Classic Cars magazine. *Owner: Captain David Wixon*

Right: In 1955 Lincoln built twelve show cars using regional themes; one was called the Sportsman and was aimed at the Southeast. The Lincolns were used on the auto show circuit for a year, and then, as was the practice, such cars were sold off through the dealer network. This one ended up in a chicken coupe many years later, rusted badly but not destroyed. Roy Tucker found it not far from his home in Massachusetts and spent the next three years blending two cars together to reclaim the 1955 Lincoln Sportsman's former glory. It is finished in sunshine yellow with maple frames and Honduras teak Formica panels. *Photo: Al Miele. Owner: Bernard Glieberman*

Ford continued as the champion of the woody wagon into the mid-fifties. This 1956 Ford Country Squire did not have any real wood in the body trim but used an ash-look fiberglass framing with walnut Di-Noc paneling. These wagons are rare today and are recently of great interest to the collectors' market. Available as a six- or eight-passenger configuration, the Country Squire could be optioned up from the stock six-cylinder to the 292ci Y-block V-8. *Owner and photo: Jim Mueller*

The 1957 Mercury Colony Park featured pillarless hardtop styling and a massive package of fiberglass ash framing over imitation mahogany paneling. The paintwork had an interesting two-tone effect with a second color applied to the coved side panel fin. Options included a 368 V-8, air conditioning, and three rows of seats.

Ford's new flagship line, the Edsel, appeared in 1958. With it came a line-up of models that included the exotic wood-trimmed Edsel Bermuda wagon. Available in both six- and eight-passenger models and powered by commanding 361 and 410ci V-8 engines, the Bermuda featured a woody insert in the lower front quarter and a large panel in the tailgate surrounded by molded fiberglass "wood" ribbing. *Photo: Phil Skinner Photo Collection*

Chapter Six

The Sixties

Once again in 1960 the new line of Ford and Mercury Country Squire and Colony Park wagons featured fake wood paneling and framing. The Ford models used boat deck paneling, while the Mercury models were fitted with simulated grained mahogany. The Mercury was an interesting design with many compounded curves, wraparound windshields, and pillarless four-door styling. Through the sixties, Ford and Mercury continued to offer Country Squires and Colony Park models trimmed in wood.

Ford added Country Squire "wood packages" to the Falcon in 1962, the new Fairlane 500 in 1963, and Mercury Comet Villager in 1964.

Chrysler got back into the woody wagon business in 1965 with a top-line version of its huge Dodge Custom 880 wagon. It featured a massive imitation mahogany panel that ran from nose to tail edged with a fine molding of chrome trim.

Buick followed suit in 1966 with a nine-passenger version of its Vista-roofed Sportwagon trimmed with simulated mahogany wood grain from the waistline down. Chevrolet also tagged along with a "wooded" Caprice wagon.

By 1967 Chevrolet, Pontiac, Jeep, and even Rambler were chasing Ford with woody wagons, adding the Chevrolet Chevelle, Pontiac Tempest, Jeep Super Wagoneer, and the Rambler Ambassador to the growing list of woody look-alikes.

Designers must have been running out of ideas when Buick introduced its Deluxe Sportwagon in 1968. This full-sized wagon featured a Vista roof and a rolling wave of simulated wood paneling that crested at the leading edge of the front door and rolled down to a thin line at the rear wheel.

Ford also exported its wood look to Australia. This 1964 XM Falcon Squire wagon was an expensive model in a country used to utility rather than style. It featured 3M Di-Noc yacht-deck planking and fiberglass ash frames. The wagons were basically the same Falcon that was sold in the United States and were powered by a 170ci six-cylinder engine. The Squire lasted three model years and is ultra rare today because the high UV content of the bright Australian sunlight destroyed the wood-grained paneling. *Photo: Ford Australia Archives*

This early sixties Morris Minor 1000 Traveller featured English white ash woodwork surrounding body-colored sheet metal, similar to a 1950 Chrysler Town and Country. Morris offered a wooded Traveller from 1954 until April 1971. They built thousands of wooded wagons, which were sold in England and Europe and exported to Africa, Australia, and America. Powered by a 1,000cc four-cylinder engine, over a million Minors, including vans, convertibles, pickups, sedans, and wagons, were assembled. *Photo: British Motor Industry Heritage Trust*

In 1969, they reversed the look and put the simulated wood at the top and colored sheet metal in the lower section!

Even though GM offered quite a variety of simulated wood-sided vehicles, Chrysler did its utmost to chase Ford's success with the Country Squire and Colony Park. Chrysler reintroduced the wood-trimmed Town and Country in 1969 fitted with a "Doorgate" rear door using the same body Dodge used for its woody Monaco model. The Doorgate was a two-way rear door tailgate that appeared on wagons from The Big Three.

Ford introduced its new full-sized line, including the Torino and the Montego, adding a wood-sided version to each. By the end of the sixties everyone was turning out a model trimmed with simulated wood. Chevrolet's El Camino could be ordered with a simulated wood panel across the tailgate, and even the funky International Harvester Travelall could be ordered "wood trimmed."

The British motor industries' biggest contribution to modern automobiles must be the Morris Mini. It was designed by Alec Issigonis, a British engineer of Turkish/Greek decent, who had designed the equally famous Morris Minor. Issigonis first proposed a front-wheel-drive minicar in 1948; however, his little dream car did not emerge until 1959 when the Morris Mini came on the market. The wagon version emerged several years later as either a miniature sedan delivery wagon or as the wood-trimmed Traveller. *Photo: British Motor Industry Heritage Trust*

This 1963 Ford Country Squire was one of 16,000 built that year. It could be optioned up with a 390ci V-8, nine-passenger seating, and a host of other comfort options. The fiberglass ash framing was rather heavy on the 1962, and with its Di-Noc mahogany paneling, it certainly stood out in the crowd. *Owner: Moon Equipment Company*

FORD COUNTRY SQUIRE WAGON

WANTED: FORD Country Squire.
DESCRIPTION: Covered-wagon loadspace–
sedan-smooth ride. Travels armed with
total performance–and can outmaneuver any posse.
Looks like a dude, but is over
200 pounds tougher
than any other wagon in its field.
Famous for quick getaways!

TRY TOTAL PERFORMANCE
FOR A CHANGE!

FORD

Falcon · Fairlane · Ford · Thunderbird

RIDE WALT DISNEY'S MAGIC SKYWAY AT THE
FORD MOTOR COMPANY'S WONDER ROTUNDA
NEW YORK WORLD'S FAIR

Available as a nine-passenger wagon in 1964, this was one
of Ford's biggest wagons of the 1960s.

1960

This 1968 Mercury is a surprisingly rare example of a late sixties wallpapered woody convertible. Mercury built ten Park Lane convertibles trimmed with "yacht-deck" paneling to celebrate the twentieth anniversary of the Mercury Sportsman. This restored Park Lane is powered by a factory 428 V-8 and carries every option available for the Park Lane that year. *Owner: Jim Ashworth*

Below, left: In 1960, Mercury offered a revised nine-passenger Colony Park four-door station wagon which sat on a 126in wheelbase chassis. This massive wagon carried its 4,558lb comfortably and the optional 430ci Lincoln V8 gave it quite zippy performance. Interestingly, its styling was carried over from the 1957 Mercury with its tall wrap-around windshield. The Colony Park wagon featured four-door, pillarless hardtop styling with a huge side panel of mahogany Di-Noc trim and fiberglass beech framing. In 1960, Ford's equivalent Country Squire wagon sold 22,237 units with a base price of $2,967. However, Mercury sold only 7,411 Colony Parks with its considerably higher $3,837 base price. *Ford Motor Company*

Below, right: For the second consecutive year, the Mercury Comet presented an entirely different appearance and major mechanical advances. This 1965 model appeared to be longer, lower, and wider although over-all dimensions changed only slightly. Performance was improved by a new 200ci seven main bearing six, a standard V8 of 289ci, and the extension of Mercury's optional three-speed automatic transmission to six cylinder models. A new electrical system included an alternator as standard equipment. Steering, ride, and comfort were improved and a wider list of options was offered. The Comet Villager featured simulated wood-grain side paneling. *Ford Motor Company*

If you researched to find the last year that Detroit featured any real kind of wood, you'd be surprised to learn that it was in the eighties. Dodge offered the Li'l Red Express Truck option on the Adventurer 150 Ram series pickup, which features an ash-trimmed stepside bed. *Owner: Mike Magda*

Chapter Seven

The Seventies and Beyond

The arrival of the seventies accelerated the wood-grained theme even further. Ford added a Squire version to its Torino Ranchero pickup and the Pinto line. Oldsmobile introduced its Custom Cruiser full-sized series. The Custom Cruiser continued in name through to the nineties.

Almost any station wagon model could then be optioned to included woody "wallpaper," as it had been coined by the used car trade. Buick's prestigious Estate Wagon, American Motor's Hornet Sportabout, and even Buick's little German import, the Opel 1900 wagon, could be optioned with wood trim.

This trend to "woodenize" every wagon spread like a new plague of pink flamingos. Chevrolet added the tiny Chevette to the list with a model known as the Woody Coupe. It was not actually a wagon but rather a hatchback that had been wood trimmed on its lower half and then upgraded with a special interior and wood-grained dash. Chevrolet even woodenized the Vega with a similar package.

Chrysler was a front runner in the woody wagon business by now. In 1973, the company offered an assortment of wooded wagons including the Plymouth Satellite Regent. This Chrysler woody-wagon look continued through the seventies; then, in 1978, the Town and Country name tag was reintroduced with the LeBaron Town and Country wagon based on the midsized Volaré/Aspen. A less ornate version of this wagon was also offered in the Aspen/Volaré line-up.

Woody kits for VW Beetles were also available from small companies. These kits used a combination of fiberglass 1940 Ford-look parts and real wood framing. The effect was quite stunning, and even ***Popular Mechanics*** picked up on the idea and offered a set of plans for the home builder.

A company called Mini-Woodie in Southern California built very stylish turnkey VW woodys and kits for a while. They were easily obtainable in the early eighties, and plans for this type of conversion still can be found in the classified section of kit car and VW magazines.

This 1978 Chrysler LeBaron Town and Country wagon is based on the Aspen/Volaré midsized model introduced in 1977. It replaced the full-sized Chrysler Town and Country wagon, which was not so heavily trimmed. This is a fully optioned special version, ordered by Ruth Ryan in New Mexico. It has a white leather interior, Police Interceptor V-8 engine, and Enforcer suspension. It was such an unusual order that Chrysler called her to make sure it was exactly what she wanted before they built the wagon. *Owner: John Ryan*

Detroit continued the woody wallpaper look in the eighties on any wagon that moved; however, Chrysler changed that idea fairly quickly when it got serious with the K-car. Lee Iacocca took a Chrysler 400, lopped off the top, and had a convertible built. It received great public reaction and went into production. The next step was to revive the great Town and Country convertible theme, and as a result, the LeBaron Town and Country convertible was introduced in 1983.

This provided Chrysler with the incentive for a woody version of the K-car Town and Country wagon in 1985. It also introduced a wallpapered version of the minivan Voyager in 1986 without any fake wood framing.

Ford had started off the decade with a full range of woody wagons, but by the late eighties it had pared the woody line down to the full-sized Country Squire and Colony Park.

Over at GM a similar chain of events occurred. The company offered both mid- and full-sized wagons trimmed with wallpaper in the line-up. Oldsmobile featured mahogany while Buick used lighter maple or birch paneling.

The Oldsmobile Custom Cruiser, Pontiac Bonneville/Parisienne, and Buick Electra Estate used the identical wagon with changes in the nose, trim, wood color, and decoration.

Outside the influence of Detroit, classic and hot rod hobby builders carried the banner for the woody theme. They lovingly called them "furniture on wheels," "rolling termite mounds," and "stick cars."

A cottage industry of woody restorers, parts suppliers, and body builders evolved as the nostalgia theme among car collectors and hot-rodders matured. Shops like the Wood'n Carr of Signal Hill, California, and Posies of Hummelstown, Pennsylvania, are restoring classic

The age of wallpapered wagons continued on through the eighties. This 1984 Oldsmobile Custom Cruiser is used by the author to tow his Hobie Cat to the lake and as a camera platform when needed. Chevrolet, Buick, Pontiac, and Oldsmobile all offered "woody" versions of this wagon during the eighties.

woodys and building a fascinating mix of high-tech hot-rod woody wagons and sedans.

As Detroit comes into the mid-nineties it continues to offer wallpapered woody station wagons: Buick with its Roadmaster Estate and Oldsmobile with its Custom Cruiser wagon and its unique Vista Window over the rear seat. Ford's woody wagon option came to an end, for now at least, in 1991 when the full-sized Squire and Colony Park wagons ceased production. However, Chrysler has kept the woody theme alive with a minivan version and a Jeep Grand Wagoneer trimmed in bodyside wood-grain appliqué.

Automotive designers like Thom Taylor have not sat still on the custom woody front. The Thom Taylor-designed 1932 woody, which Dan Fink built for himself, is a stunning example of what can be done when superb design meets great craftsmanship. Thom has also designed the new 1937 Ford Extremeliner for Posies, the famous hot rod builder mentioned above. The world has not seen the last of the woody. Several aftermarket firms are now offering real wood kits for Jeep Grand Wagoneers and Ford Explorers. The idea of adding pizzazz to a vehicle with trim and wood has been a public and corporate obsession since the beginning of the automobile, and these two themes have survived to this day.

Detroit may never produce a real woody again—in these modern times it would be too labor intensive to be cost effective. However, with limited model runs like the Dodge Viper demonstrating a bright economic forecast, you never know what style might lurk in the hearts of young designers, shaping tomorrow's transportation fun.

MICHIGAN
073 YHT
GREAT LAKES

1GUD345

Left top: The 1994 Buick Roadmaster continued to offer wood-grained wallpaper as part of its wagon package. The Roadmaster wagon comes with a vista roof panel over the back seat, a nine-seat interior, and alloy wheels.

Left bottom: In 1983, Chrysler announced the new Town and Country convertible based on the Dodge CV 400. It met with great success. The year also had other great highlights for Chrysler. The company paid back a massive loan to the government years ahead of schedule, much to the surprise of the government and other manufacturers who had seen only gloom hanging over the auto industry.

The Mill-Custom Mill Products from Perrysburg, Ohio, have revived the real wood trim idea in the '90s with the "Real Wood" conversion for Jeep Cherokees and the new Grand Cherokee. Made with walnut panels and white ash framing, the conversion adds a little snap to the Jeep's off-road character. *Photo: The Mill-Custom Mill Products*

IS NOT LIABLE
FOR FIRE, THEFT,
PARTS LEFT
MORE THAN 24
MONTHS.

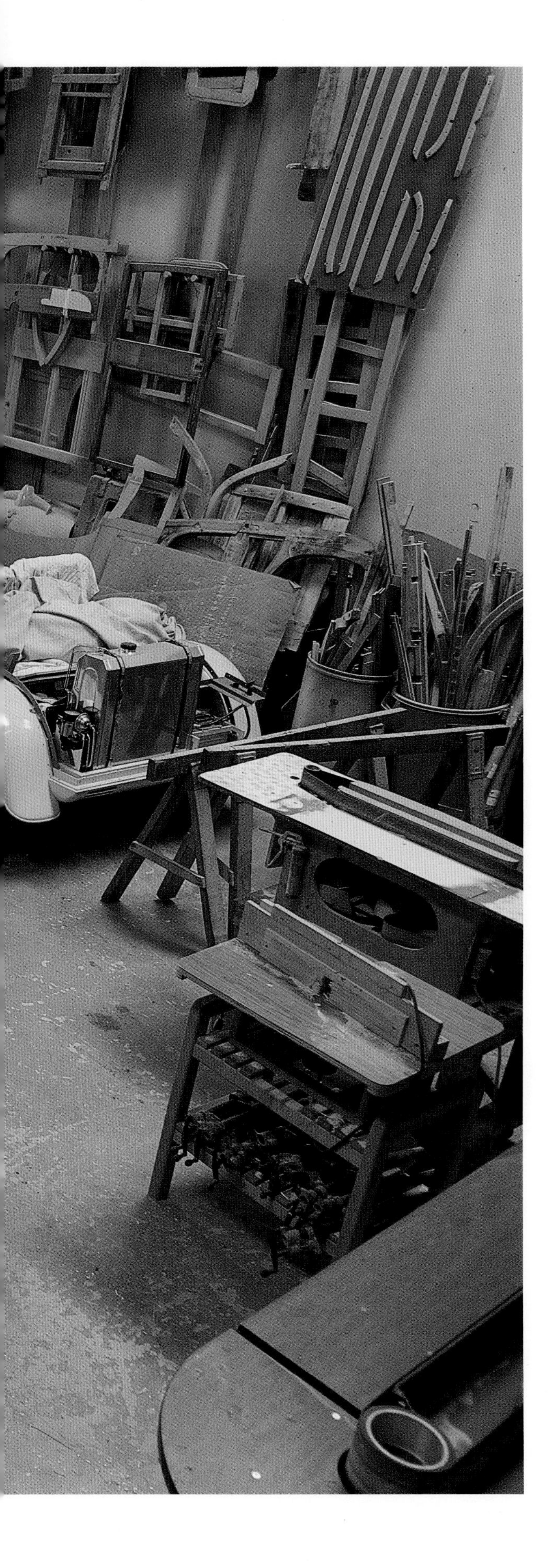

Chapter Eight

Woody Restoration and Care

Restoration of any woody is a time-consuming and expensive proposition. Doing it yourself requires more than just a working knowledge of wood, hammers, saws, and nails. Many folks have done their own work with great results, but getting there is often quite a task.

Early depot hack-style wagons are the easiest to build as they have few, if any, curved pieces to replicate; however, a Town and Country body has more curves than a stage full of Rockettes and requires the talents of a master wood craftsman to do the work. Full rebuilding requires hundreds of labor-intensive hours. Assembly is extremely intricate with complex finger joints and compound curves that not only wrap around and down the trunk line but roll around their own vertical axes.

There are some well-documented books on the subject of building wood-structured vehicles if you intend to tackle your own project, although thorough maintenance is critical once it is restored.

Wood Care

As you can see from some of the vehicles we have included in this book, it is possible to keep wood in great shape for decades if you store it dry and treat it with respect. Some 1930s wagons we featured still have all their original wood.

For best results, don't drive it in the rain or on wet pavement; after all, you don't put your household

A woody restoration shop like The Wood'n Carr looks like a regular carpentry shop; however, the talent here is so specialized that no ordinary carpentry shop could hope to achieve the level of fit, form, and perfection that Doug and Suzy Carr roll out through the door. Across the country there are only a handful of such shops that specialize in wood wagon restorations. Doug and Suzy's stunning black 1934 Ford station wagon can be seen parked just inside the door. The woody was the center spread feature in Hot Rod magazine's swim suit edition several years back.

At The Wood'n Carr in Signal Hill, California, Doug Carr is seen here working on a "phantom" 1947 two-door Ford station wagon, which he fabricated out of eastern hardrock maple. A full restoration or custom-built wagon like this generally takes about six months at The Wood'n Carr shop, but the results are jewel-like.

Various framing and carpentry methods are used in almost every facet of a full wood reconstruction. Some sections are molded with ribbing while other parts feature heavyweight framing. Restoring or building a wagon frame is a time-consuming process, even with a highly mechanized shop. Every part still has to be handmade, then hand-fitted, pre-assembled, assembled as a completed body, sanded, and finally finished off with varnish. Doug Carr still prefers to use varnish as it brings out the wood's golden colors without having to stain it. Doug likes to finish a new body with as many as thirty coats of varnish and then rub it to a high-gloss finish.

Anyone who has serviced or built a wood-sided vehicle knows about sanding. Due to the shape and contours of the wood, most sanding can only be done by hand, and to get the quality of finish demanded by restorers today, hundreds of hours of hand-sanding go into a single wagon body.

furniture out in the rain. The same applies to your woody.

If you need to completely refinish an already restored vehicle, you have two choices: either have it done professionally, or spend the time helping out on other woodys so you can learn to do it correctly before refinishing your own. Chemical strippers are not recommended as the liquid can stain the wood and cause it to swell unnecessarily. Stripping the varnish off the framing and paneling should be tackled with 180- or 280-grit sandpaper, taking it down to bare wood.

Stains from water and mildew in the timber framing and paneling can be removed with oxalic acid-based wood bleaches, which should be neutralized after several hours, washed, and allowed to dry. A wood preservative treatment should then be applied and allowed to dry.

The wood must be sanded once again with 320- and then 400-grit paper and gently blown free of dust so as not to raise fibers on the wood surface. The first coat of varnish should be thinned according to instructions as this allows the varnish to better penetrate the wood surface and create a good base for the subsequent coats. This coat should then be lightly scuffed and cleaned with a fresh tack rag.

After each coat dries, subsequent coats should be applied without scuffing. The number of coats is up to the owner. Original owner's manuals noted a minimum of two coats, but many more is a prudent way of getting a quality, long-lasting finish.

Surprisingly, Doug Carr, the owner of The Wood'n Carr, still recommends finishing and refinishing woodwork using traditional varnish. Doug feels this has a twofold effect: It gives the wood its golden luster, allowing it to gracefully age to the correct tone, and it provides excellent sealing of the wood so no moisture can penetrate it. Doug has been known to apply up to thirty coats before he's done.

Regular inspection of all joint and mating surfaces should be done at every cleaning. Any joints that have opened up should be attended to immediately so no water can penetrate them. If a joint has cracked or opened up, it should be cleaned out, repaired, and refinished quickly according to the product manufacturer's instructions.

Cleaning of wood surfaces should be done with a damp terry cloth, rag, or diaper and dried with a fresh cotton towel or a chamois. Doug Carr emphasizes that wood bodies should not be washed, ever! A final air dry in late afternoon sun is also suggested. Waxing is highly recommended using a cleaner wax to help the wood resist road dirt and tar.

Storage of wooden-bodied vehicles is critical. The storage area should be dry, well ventilated, clean, and free of direct sunlight to best maintain your woody. If you live in a harsh climate, your garage should maintain a constant temperature and humidity level.

These few tips on care and preservation of your woody will give you an idea of how much work woodys require to keep their glow. Remember, the final quality of any finishing job is totally dependent on how well you do the preparation and varnish work.

Finger joints like this one used on the Town and Country are extremely difficult pieces of carpentry but very necessary in any T&C wood restoration.

Chapter Nine

Odd and Interesting Woodys

Like many different fields of automotive endeavor, the woody has had its share of dreamers, imitators and craftsmen who have created their personal visions of wood on wheels.

In the beginning, the theme and creation of the woody depot hack was almost always a custom build and design, but as the automobile evolved and flourished it became both a working production vehicle and an exclusive custom built automobile.

Seen in this chapter are some very innovative custom woodys that were built by folks who loved the woody concept and the lifestyle it portrayed.

George Barris's "Surf Woody" was displayed across the cover of ***Hot Rod*** magazine in the mid sixties and was a hit of the custom car show circuit for years. Due to the continuing interest in this particular woody it has recently been restored.

Even the little VW Beetle got the fully woody treatment. ***Popular Mechanics*** magazine offered a set of plans to build-your-own in the early seventies and in the late seventies in Southern California Paul Wilson built a short run of stunning Mini-Woodie conversions. Modern day hot rod designers like Thom Taylor have generated an interest in woodys by creating unique phantom woodys which builders like Posies, Doug Carr at The Wood'n Carr," Boyd Coddington, and Dan Fink have taken to heart and built.

We have not seen the last of the woody! Without a doubt, each summer somebody is going to roll out yet another restoration, conversion, phantom or wild woody.

This photo from the editor of France's best hot rodding magazine Nitro, is of fellow Frenchman, Sylvain Laurensot's brilliant yellow woody wagon built using a mixture of classic Citroen 2CV components. Sylvain's tiny front-wheel-drive wagon features a custom steel flip front from a 1955 model, the chassis and drive train from 1966 Ami 8 and a hand-fabricated woody body built which was built over a tubular steel structure using beech frames and oak-tinted marine ply for the paneling. *Claude Lefebvre.*

The Surf Woody was one of George Barris' most interesting projects in the mid-1960s. With the help of engineer Dick Dean, the Surf Woody was built on a tubular steel frame and came with a removable back section that made the wagon into a roadster. Powered by a twin supercharged Ansen-built Ford V-8, it also featured a roll-around fluorescent headlight assembly, a Muntz stereo system, TV, and an antique French telephone. The body was trimmed in carefully matched African mahogany veneer over candy orange bodywork. *Owner/builder: George Barris*

Above: The Calico Surfer Woody was a George Barris-designed woody for the seventies. It featured a tube frame using Corvair suspension and a Ford rear end and was powered by a supercharged Mustang 289 V-8. Built with the help of Dick Dean, the Calico Surfer was framed in ash and paneled in mahogany that had been stained with a blue tint to give it the look of blue calico fabric. *Owner/builder: George Barris*

Below: This snappy-looking little Model A Ford woody wagon replica was built by its owner several years ago based on Chevette components using a kit offered by Legendary Motorcars from Columbus, Ohio. It features a custom steel chassis, four-cylinder Chevette engine, automatic transmission, and seating for four. *Owner: Ray Toronto*

The "Mini-Woodie" was designed and built by industrial designer Paul Wilson. Its shapes, colors, and textures blend nicely with the deco lines of the VW, creating a sweet little woody. Paul uses it as a daily driver most days in sunny Southern California. *Owner: Paul Wilson*

The latest generation of hot rod woody wagons from Thom Taylor is currently being built by Posies in Hummelstown, Pennsylvania. Known as the Extremeliner, this totally hand-fabricated hot rod is slated to appear sometime in 1995. With hot rods like this, it seems that the future is already here, and if designers like Thom Taylor continue to push their creativity to the max, then we can rest assured that the woody will not only live in our hearts but continue to cruise into the next century. *Illustration: Thom Taylor*

Reading and Reference List

Chrysler's Wonderful Woodie: The Town and Country. Donald J. Narus. Venture Publishing, Parma, OH, 1988.

Encyclopedia of American Cars 1940-1970. Richard, Langworth. Beckman House, New York, NY, 1980.

Fords Forever. Lorin Sorensen. Motorbooks International, Osceola, WI, 1990.

Ford 1903 to 1984. Lorin Sorensen, D. Lewis, and M. McCarville. Beekman House, New York, NY, 1983.

Great American Woodies and Wagons. Don Narus. Crestline Publishing, Glen Ellyn, IL, 1977.

Special Interest Auto Magazine.

The Buick: A Complete History. Terry B. Dunham and Lawrence R. Gustin. Automobile Quarterly Publishing, Kutztown, PA, 1987.

The Complete History of Chrysler. Richard Langworth, and Jan Norbye. Beekman House, New York, NY, 1985.

The Packard Story. Robert Turnquist. A.S. Barnes & Co., New York, NY, 1965.

The Station Wagon: Its Saga and Development. Bruce Briggs. Vantage Press, New York, NY, 1975.

'32 Ford—The Deuce. Tony Thacker. Osprey Automotive, London, Great Britain,1984.

Woodie Times

Woodie Woodworking. Rich Bloechl. Bay Area Graphics, San Jose, CA, 1993.

Index